SpringerBriefs in Mathematical Physics

Volume 29

SpringerBriefs are characterized in general by their size (50-125 pages) and fast production time (2-3 months compared to 6 months for a monograph).
Briefs are available in print but are intended as a primarily electronic publication to be included in Springer's e-book package.

Typical works might include:

- An extended survey of a field
- A link between new research papers published in journal articles
- A presentation of core concepts that doctoral students must understand in order to make independent contributions
- Lecture notes making a specialist topic accessible for non-specialist readers.

SpringerBriefs in Mathematical Physics showcase, in a compact format, topics of current relevance in the field of mathematical physics. Published titles will encompass all areas of theoretical and mathematical physics. This series is intended for mathematicians, physicists, and other scientists, as well as doctoral students in related areas.

More information about this series at http://www.springer.com/series/11953

Masao Jinzenji

Classical Mirror Symmetry

Springer

Masao Jinzenji
Department of Mathematics
Hokkaido University
Sapporo, Hokkaido
Japan

ISSN 2197-1757 ISSN 2197-1765 (electronic)
SpringerBriefs in Mathematical Physics
ISBN 978-981-13-0055-4 ISBN 978-981-13-0056-1 (eBook)
https://doi.org/10.1007/978-981-13-0056-1

Library of Congress Control Number: 2018938385

Printed on acid-free paper

This Springer imprint is published by the registered company Springer Nature Singapore Pte Ltd.
part of Springer Nature
The registered company address is: 152 Beach Road, #21-01/04 Gateway East, Singapore 189721, Singapore

Preface

This book is intended as a brief introduction to classical mirror symmetry. I mean by the term "Classical Mirror Symmetry" the process of computing Gromov–Witten invariants of a Calabi–Yau threefold by using the Picard Fuchs differential equation of period integrals of its mirror Calabi–Yau threefold. As history tells us, the study of mirror symmetry has developed explosively both in physics and mathematics since the 1990s. In physics, it caused the discovery of "string duality" and the second revolution of string theory. This movement evolved into the study of D-branes in string theory, which are still major topics among string theorists. In mathematics, it caused rigorous construction of moduli spaces of stable maps and Gromov–Witten invariants. Nowadays, this direction is evolving into rigorous construction of open Gromov–Witten invariants and categorical mirror symmetry, which are also affected by the study of D-branes. On the contrary, this book aims at a brief exposition of the study of mirror symmetry at its early stage. Therefore, I concentrate on the best known example, the quintic hypersurface in four-dimensional projective space and its mirror manifold. Of course, I also explain my recent results on classical mirror symmetry in the last part of this book. The style of description is just between that of mathematics and physics. I assume that the readers have standard backgrounds of both physics and mathematics that are expected for graduate students.

This book is organized as follows.

In Chap. 1, I briefly review the process of discovery of mirror symmetry and the striking result proposed in the celebrated paper by Candelas and his collaborators.

In Chap. 2, I explain some elementary results of complex manifolds and Chern classes needed for study of mirror symmetry.

In Chap. 3, I introduce topological sigma models, A-model and B-model, and explain the geometrical meaning of correlation function of these models.

In Chap. 4, I explain basic results of toric geometry and the process of construction of a mirror pair of Calabi–Yau threefold by using toric geometry. Then, I demonstrate computation of the A-model Yukawa coupling of the quintic hypersurface under assumption of the mirror symmetry hypothesis (so-called B-model computation).

In Chap. 5, I reconstruct the B-model computation of Chap. 4 from the point of view of the moduli space of holomorphic maps (A-model), which corresponds to a summary of my recent works.

Lastly, I would like to thank Prof. T. Eguchi and Dr. H. Saito for valuable discussions. I would also like to thank Ms. C. Hasebe for editorial support and my family for warm support during the writing period of this book.

Sapporo, Japan Masao Jinzenji

Contents

Chapter 1
Brief History of Classical Mirror Symmetry

Abstract In the background to the discovery of the mirror symmetry hypothesis there exists an idea of compactifying extra 6 dimensions of 10-dimensional heterotic string theory in order to obtain a 4-dimensional grand unified theory that describes our real world. The complex 3-dimensional Calabi–Yau manifold is nothing but the compact 6-dimensional space used for this purpose. In this chapter, we explain why this idea came from particle physicists and give a brief exposition of the process of discovery of mirror symmetry. Lastly, we introduce the first striking result of classical mirror symmetry revealed by the celebrated paper by Candelas et al. (For details of compactification of string theory, we recommend readers to consult [1].)

1.1 Grand Unified Theory and Superstring Theory

Since the striking paper, "A Pair of Calabi–Yau Manifolds as an Exactly Solvable Superconformal Field Theory" [1] by Candelas, de la Ossa, Greene and Parks, which appeared in 1991, mirror symmetry has attracted the interest of not only particle physicists, but also a large number of mathematicians. The main result presented in the paper, which corresponds to a "Conjecture" in the sense of mathematics, suggests that a difficult problem in algebraic geometry, "counting rational curves in a quintic hypersuface in complex 4-dimensional projective space", can be solved by analysis of a linear differential equation, which seems to be unrelated to the original problem at first sight. Moreover, the process of analysis of the differential equation can be interpreted as an extension of obtaining Fourier coefficients of modular forms by using linear differential equations for period integrals of a family of elliptic curves. Hence the conjecture is quite attractive to mathematicians, but behind its discovery lies an idea of constructing a grand unified theory (GUT) from superstring theory, which originated from particle physicists. Let us explain the outline of this idea.

The grand unified theory aims at unifying gravitational, electromagnetic, weak and strong interactions, that are observed in our real world, in a single framework. Why do we have to unify them? It is indeed a difficult question, but theoretical physicists think that the number of fundamental laws in physics, which corresponds to axioms in mathematics determined from observation of the real world, should

M. Jinzenji, *Classical Mirror Symmetry*, SpringerBriefs in Mathematical Physics, https://doi.org/10.1007/978-981-13-0056-1_1

be smaller. Attempts at a unified theory began in the era of Einstein. In the first place, Einstein discovered his special relativity by pursuing consistency between electromagnetism and the Galilean principle of relativity in Newtonian mechanics, and then constructed his general relativity by pursuing consistency between special relativity and the classical theory of gravity. His motivation for the celebrated achievements was consistency of fundamental laws in physics. Therefore, it was natural for him to aim at unifying interactions known in his era, electromagnetic interaction and gravitational interaction. Some other physicists in his era also participated in this project. Let us introduce "Kaluza–Klein theory", which is one of the celebrated theories in attempts at unification. This theory assumes that our world is not $1+3=4$-dimensional but 5-dimensional with an extra dimension. This extra dimension corresponds to very small circle (S^1), invisible to the naked eye. Let us illustrate this idea by considering 1-dimensional objects with an extra dimension that corresponds to a small S^1. It looks like a 1-dimensional string to the naked eye, but turns out to be a hollow tube when viewed using a microscope. If we apply general relativity theory to this 5-dimensional world, among elements of Riemannian metric $g_{\mu\nu}$ $(\mu, \nu = 1, 2, 3, 4, 5)$, $g_{\mu 5} = g_{5\mu}$ $(\mu = 1, 2, 3, 4)$ can be interpreted as the vector potential A_μ in electromagnetism (since the extra dimension is invisible). If we restrict a 5-dimensional Einstein equation of gravity to this part, we can derive Maxwell's equation of electromagnetism. This is the outline of Kaluza–Klein theory. This theory is not a quantum theory but a classical theory and looks naive from a modern point of view. But the idea of compactification of string theory originates from the theory.

As particle physics advanced from the era of Einstein, "weak interaction" and "strong interaction" were formulated in addition to the two old interactions. Weak interaction describes β decay of neutrons, for example, and was formulated by Fermi, who discovered the neutrino. Strong interaction is the interaction that glues quarks into hadrons, such as protons and neutrons. These interactions are formulated as gauge theories described by vector potential A_μ in the same way as electromagnetism. But in contrast to electromagnetism, the vector potentials take matrix values and the corresponding gauge theories are called "non-commutative gauge theory". Since electromagnetic, weak and strong interactions are formulated as gauge theory, one is tempted to unify these interactions into a framework of one gauge theory with a large gauge group. The most successful achievement in this direction is the "Weinberg–Salam model" that unifies electromagnetic and weak interactions. In modern physics, it is called the "standard model" and has been confirmed by many experiments in particle physics. Later, $SU(5)$ gauge theory, which unifies the two interactions with strong interaction, was proposed, but it predicts proton decay, which has not been observed by experiments. Hence it has not been confirmed. Gauge theory that unifies the three interactions has been studied up to the present day, but a theory that takes the place of the Weinberg–Salam model has not appeared yet.

On the other hand, attempting to unify gravitational interaction with other interactions, which was done at the earliest stage of unified theory, had turned out to be the most difficult problem as quantum field theory advanced. Of course, we have difficulty coming from the fact that general relativity is described by Riemannian metrics $g_{\mu\nu}$, which have different characteristics from gauge field A_μ. But we can overcome this difficulty by using the idea of Kaluza–Klein theory at a classical level. The most serious difficulty comes from the fact that general relativity is non–renormalizable as a quantum field theory. It is caused by high non-linearity of general relativity compared with gauge theory. If we try to quantize general relativity by using a standard recipe of quantum field theory, we have to add an infinite number of counter terms to the Lagrangian in order to cancel too many divergences caused by high non-linearity. In this way, attempts at a grand unified theory confronted the difficulty of quantization of general relativity, which seems to be insurmountable wall.

In the 1980s, superstring theory came into the spotlight as the theory that we can expect to overcome all the difficulties of a grand unified theory. This theory originates from the dual resonance model, proposed by Veneziano in 1960s, which describes interaction of hadrons. Nambu and Susskind pointed out that the mathematical structure of this theory is well understood under the assumption that the fundamental object of the theory is not point particles but 1-dimensional strings. In this way, string theory was studied at its early stage as a theory that describes the mechanics of hadrons. But it turned out that the theory predicts an unwanted spin 2 particle (tensor field of rank 2, $A_{\mu\nu}$), and physicists at that time gave up the study of string theory. In the early 1980s, Green and Schwarz were studying vector bundlestring theory, which is string theory with vector bundlesymmetry.

Supersymmetry is a symmetry which assumes that bosons and fermions together form pairs in quantum field theory. Roughly speaking, a boson is a particle with integer spin, such as gauge particles that mediate interactions, and it is described by a field which takes the value of a commutative real number. On the other hand, a fermion is a particle with half integer spin, such as quarks and electrons that correspond to matter, and it is described by a field which takes the value of an anti-commutative grassmann number. Supersymmetry is formulated as a symmetry that exchanges the commutative field and the anti-commutative field. If supersymmetry is added to a quantum field theory, the theory acquires good characteristics such as having less divergent amplitudes, etc. Therefore, many supersymmetric versions of known theories were proposed and studied. Superstring theory appeared in this trend.

In studying string theory, we have to consider a "world sheet", which is a real 2-dimensional surface given as the trajectory of motion of 1-dimensional string. Therefore, studying the mechanics of a string is nothing but studying the mechanics of a map $\phi : \Sigma \to M$ from a 2-dimensional surface Σ to a space-time manifold M. Nambu and Goto constructed the first fundamental action of string theory (Nambu–Goto action) by using the idea that the area of the world sheet $\phi(\Sigma)$ in M should be minimized. But this action contains a square root and has highly non-linear characteristics. Hence quantization of the action is difficult. Therefore, physicists began to use Polyakov action, which gives the same classical motion of strings as Nambu–Goto action but has linear characteristics. In order to quantize string theory, Polyakov used the trick of introducing an auxiliary Riemanniann metric $h_{\mu\nu}(x)$ of Σ itself. Polyakov action has the notable feature of "conformal invariance". This invariance means that the action is invariant under conformal transformation $h_{\mu\nu}(x) \to e^{\varphi(x)} h_{\mu\nu}(x)$.

This invariance plays a very important role in considering quantization of string theory. Of course, superstring theory was constructed by adding a fermion field ψ as superpartner of a boson field ϕ, to Polyakov action.

In quantizing string theory and superstring theory, we have possibility of breakdown of conformal invariance caused by quantum anomaly terms. It turns out that the conformal anomaly term vanishes if space–time M is 26-dimensional in the case of bosonic string theory and 10-dimensional in the case of superstring theory respectively. Therefore, we have to consider space-time whose dimension surpasses four in order to construct meaningful physics from string theory.

One reason why superstring theory became a candidate for the grand unified theory comes from the fact that it contains a spin 2 tensor field. If we regard this particle as the Riemannian metric $g_{\mu\nu}$ of general relativity, this feature once considered as weak point becomes a strong point of string theory. As was mentioned before, quantization of general relativity has a serious divergence problem. But if we assume that general relativity is induced from string theory, we can expect a solution of this problem. Divergence of quantum field theory occurs from loop Feynmann diagrams in perturbation calculation. In the case of perturbation calculation of string theory, the loop diagram is replaced by a closed 2-dimensional surface with holes (in other words, with positive genus). Because of conformal invariance, this surface is identified with a 1-dimensional complex manifold, i.e., Riemann surface. Therefore, the amplitude of the loop diagram is replaced by integration on the Riemann surface with positive genus in the case of string theory. The theory of complex analysis tells us that integration on Riemann surface with positive genus has good characteristics called "modular invariance". This invariance suppresses divergence of loop amplitudes. Hence superstring theory has attractive feature of quantization of general relativity.

Let us go back to Green and Schwarz. They were studying a version of superstring theory called Type I theory. It contains both a closed string homeomorphic to S^1 and an open string homeomorphic to the closed interval $[0, 1]$. Since the charge of the gauge field is attached to both ends of the open string, the theory includes gauge theory. In quantizing this theory, various quantum anomaly terms that break gauge symmetry appear. But they found out that all these terms vanish if they take the Lie group $SO(32)$ as gauge group. This discovery tells us the following. This theory includes both general relativity and gauge theory and a physically meaningful gauge group is uniquely determined from intrinsic reasoning. In constructing a unified gauge theory that contains electromagnetic, weak and strong interactions, we had no intrinsic motivation for the choice of gauge group other than the condition that it contains respective gauge groups as its subgroups. Therefore, their discovery was very attractive to many particle physicists at that time. In this way, a boom called the "first string theory revolution" occurred in the early 1980s.

In this movement, several versions of superstring theory were proposed as candidates for a grand unified theory. The version that Candelas et al. used as the background to their paper is called "heterotic string theory". In quantizing string theory, we first solve the classical equation of motion of string derived from Polyakov action. In this process, we obtain two types of solutions called the left moving solution (holomorphic solution) and the right moving solution (anti-holomorphic solution) respectively. Quantization can be done independently for these two solutions and the corresponding theories are called the left moving theory and the right moving theory respectively. Heterotic string theory adopts 26-dimensional bosonic closed string theory as its left moving theory and 10-dimensional closed superstring theory as its right moving theory. The extra 16 dimensions of the left moving theory are compactified by a 16-dimensional torus ($(S^1)^{16}$), which produces a gauge group of rank 16. If we make the gauge group into $E_8 \times E_8$ or $SO(32)$ by tuning the shape of the 16-dimensional torus, we can obtain 10-dimensional anomaly-free string theory in the same way as Type I theory. In the classical limit, this theory turns out to be 10-dimensional supergravity theory with gauge symmetry, but what we need is a 4-dimensional grand unified theory. Therefore, we have to compactify the extra 6 dimensions by using a real 6-dimensional compact manifold. This 6-dimensional manifold is nothing but the main stage of classical mirror symmetry conjecture proposed in the paper by Candelas et al. We impose the condition that the resulting 4-dimensional theory has supersymmetry. It follows that the 6-dimensional compact manifold must have an $SU(3)$ holonomy group. In other words, the manifold must satisfy a kind of flatness condition, that is, having "everywhere vanishing Ricci curvature". The 6-dimensional manifold that satisfies this condition is called a "complex 3-dimensional Calabi–Yau manifold", which is a major subject in mathematics. In this way, interests of particle physicists and interests of mathematicians began to overlap.

1.2 Compactification of Heterotic String Theory

In the previous section, we introduced the scenario of obtaining 4-dimensional grand unified theory (GUT) from compactifying 10-dimensional $E_8 \times E_8$ heterotic string theory by a 6-dimensional compact manifold K. Let assume first that the radius of the closed string is of Planck scale order ($\simeq 10^{-35}$ m). It follows that particles that correspond to excitation modes of the string have extremely large mass and cannot be observed in the real actual world. Therefore, we only consider particles that correspond to massless modes of the string (we call these particles "massless particles"). This fact reduces 10-dimensional heterotic string theory into its effective low energy field theory, i.e., 10-dimensional supergravity theory. The supergravity theory that corresponds to $E_8 \times E_8$ is given by $N = 1$ supergravity theory coupled with $E_8 \times E_8$ Yang–Mills theory. Let us introduce here subscripts to represent fields in this theory. We use $M, N, \ldots$ for subscripts of 10-dimensional space, $a, b, \ldots$ for subscripts for representation of the gauge group (in our case, adjoint representation is used), $\alpha, \beta, \ldots$ for subscripts for spinor representation of the Lorentz group $SO(1, 9)$ and $A, B, \ldots$ for subscripts for fundamental representation of $SO(1, 9)$. In the supergravity theory, the following fields appear: vielbein e^A_M $(e^A_M e^B_N \eta_{AB} = g_{MN})$ which represents the 10-dimensional metric g_{MN}, ψ^α_M which is the superpartner of e^A_M, dilaton field ϕ, λ^α which is the superpartner of ϕ, second order antisymmetric tensor field B_{MN}, Yang–Mills gauge field A^a_M and χ^a_α which is the superpartner of A^a_M. Among these fields, e^A_M, B_{MN}, A^a_M and ϕ are bosons and the remaining fields are fermions. In the theory, e^A_M, B_{MN} and A^a_M are potential fields and the actual field strength of these fields is represented by their differentials. As for the vielbein e^A_M which represents gravity, field strength is represented by curvature $R = d\omega + \omega \wedge \omega$, where $\omega = \omega^A_{MB} dx^M$ is the spin connection obtained from the differential of e^A_M. Field strength of the gauge field A^a_M is given by curvature $F = dA + A \wedge A$, where $A = A^a_M T_a dx^M$ and T_a is the representation matrix of adjoint representation. Lastly, field strength of B_{MN} is represented by the field

$$H = H_{MNP} dx^M \wedge dx^N \wedge dx^P = dB + \omega_{3L} - \omega_{3Y},$$

where $B = B_{MN} dx^M \wedge dx^N \cdot \omega_{3L}$ and ω_{3Y} are Chern–Simons terms with respect to gravity fields and gauge fields, that are introduced from discussion of anomaly cancellation of $E_8 \times E_8$ heterotic string theory. These terms are defined by,

$$d\omega_{3L} = \mathrm{tr}(R \wedge R), \quad d\omega_{3Y} = \mathrm{tr}(F \wedge F).$$

Of course, there exists the full Lagrangian of this supergravity theory, but we don't write it down here because it is too lengthy.

Next, we discuss 4-dimensional theory obtained from compactifying the 10-dimensional supergravity theory by a 6-dimensional compact manifold K. We demand that the 4-dimensional theory have supersymmetry. Let us explain briefly the reason for this demand. In our scenario, the radius of K is taken to be of Planck scale order. Therefore, in considering quantum theory of these fields restricted to the manifold K, only the zero modes of the quantum theory survive in the 4-dimensional effective theory. Because non–zero modes have such large mass proportional to the inverse of the Planck scale, that they cannot be detected on the scale of our real world. If we don't impose supersymmetry, the fields in the 4-dimensional theory are zero modes of the quantum theory on K that correspond to the fields in the 10-dimensional supergravity theory. On the other hand, in the standard model which is the most reliable model of our 4-dimensional world, there exists a Higgs scaler boson doublet of the electroweak gauge group $SU(2) \times U(1)$. In our scenario, the electroweak gauge group should be a subgroup of $E_8 \times E_8$, and only the bosonic field with subscripts of representation of the gauge group is A^a_M. But A^a_M is not a scaler field, which causes a contradiction. But if we assume supersymmetry of the 4-dimensional theory, we can construct a scaler boson field as superpartner of the field that corresponds to spin $\frac{1}{2}$ fermion χ^a_α.

Let us discuss conditions imposed on the manifold K from the assumption of supersymmetry of the 4-dimensional theory. Let Q be the generator of the supersymmetry. It follows from our assumption that 4-dimensional vacuum state $|\Omega\rangle$ must satisfy $Q|\Omega\rangle = \langle\Omega|Q = 0$. Hence for any operator U, the relation $\langle\Omega|QU + UQ|\Omega\rangle = \langle\Omega|\{Q, U\}|\Omega\rangle = 0$ must hold. If U is bosonic, $\{Q, U\}$ is fermionic and the relation holds trivially. If U is fermionic, $\{Q, U\}$ is bosonic and the relation becomes a non trivial condition. If we write the supersymmetry transformation of U as δU, $\{Q, U\}$ is nothing but δU by definition of Q. Hence the above condition is rewritten as $\langle\Omega|\delta U|\Omega\rangle = 0$, which means $\delta U = 0$ in the classical limit. In other words, the supersymmetry transformation of every fermion field must be zero in the classical limit.

At this stage, let us write down the supersymmetry transformation of fermion fields, ψ_M, χ^a and λ of the supergravity theory (we omit spinor subscripts). They are written down by using an infinitesimal bosonic spinor η as follows.

$$\delta\psi_M = \frac{1}{\kappa}D_M\eta + \frac{\kappa}{32g^2\phi}(\Gamma_M^{NPQ} - 9\delta_M^N\Gamma^{PQ})\eta H_{NPQ} + (\text{fermions})^2,$$

$$\delta\chi^a = -\frac{1}{4g\sqrt{\phi}}\Gamma^{MN}F^a_{MN}\eta + (\text{fermions})^2,$$

$$\delta\lambda = -\frac{1}{\sqrt{2}\phi}(\Gamma^M\partial\phi)\eta + \frac{\kappa}{8\sqrt{2}g^2\phi}\Gamma^{MNP}\eta H_{MNP} + (\text{fermions})^2.$$

Here, κ and g are coupling constants of the theory and Γ_M^{NPQ} etc. are Dirac gamma matrices of 10 dimensions. We tentatively ignore the terms $(\text{fermions})^2$ and consider the condition that $\delta\psi_M = \delta\chi^a = \delta\lambda = 0$ holds. For simplicity, we assume that the dilaton field ϕ is constant. Then we can see that the following conditions are necessary and sufficient:

$$D_M\eta = 0, \quad \Gamma^{MN}F^a_{MN}\eta = 0, \quad H_{MNP} = 0.$$

In order to satisfy the first condition on K, the holonomy group of K has to be $SU(3)$. Let us explain the reason in the following. First, we introduce the notion of holonomy group of K. Let us take a tangent vector $\mathbf{v}$ on a point $p \in K$. Then we consider a closed loop C in K that starts from and ends at p. If we apply parallel transport along C in the sense of riemannian geometry to the vector $\mathbf{v}$, we obtain another tangent vector $\mathbf{v}'$ on p. Since K is not a flat manifold in general, $\mathbf{v}'$ does not always coincide with $\mathbf{v}$ and is represented as $\mathbf{v}' = U(C)\mathbf{v}$, where $U(C)$ is a matrix that belongs to $SO(6)$. If we vary the closed loop C in K with starting and ending point p fixed, the corresponding matrix $U(C)$ forms a subgroup of $SO(6)$. This subgroup is the holonomy group of K. Of course, there are cases where the holonomy group coincides with $SO(6)$. But if the holonomy group is smaller than $SO(6)$, there exists some limitation on how the Riemannian manifold K is curved. At this stage, let us take note of the condition $D_M\eta = 0$. If it is restricted to K, it reduces to the condition $D_i\eta = 0$, where i is the subscript for local coordinates of K. η is a spinor of 8-dimensional representation of $SO(1,9)$, and D_i is the covariant derivative with respect to this representation. Hence the condition $D_i\eta = 0$ means that η coincides with itself after parallel transport along any closed loop in K. Let us determine the condition that such η exists. As is well-known in mathematics, the Lie algebra of $SO(6)$ (infinitesimal transformation) is isomorphic to that of $SU(4)$ (in fact, the dimensions of these groups are both 15). Hence we replace the group $SO(6)$ by $SU(4)$. Then the 8-dimensional spinor representation is decomposed into $\mathbf{4} \oplus \bar{\mathbf{4}}$ representation of $SU(4)$. Here, $\mathbf{4}$ is the fundamental representation of $SU(4)$

and $\overline{\mathbf{4}}$ is its complex conjugate. Let us assume that there exists at least one η and that its value at p lies in the representation $\mathbf{4}$. Then the value of η at p is represented as follows:

$$\begin{pmatrix} 0 \\ 0 \\ 0 \\ \eta_0 \end{pmatrix}.$$

The condition that η is invariant under parallel transport along any closed loop C is equivalent to the condition that the corresponding matrix $U(C)$ in $SU(4)$ takes the following form:

$$U(C) = \begin{pmatrix} G & \mathbf{0} \\ {}^t\mathbf{0} & 1 \end{pmatrix}, \quad (G \in SU(3)),$$

which means that the holonomy group of K is contained in $SU(3)$, which is a subgroup of $SU(4)$.

Next, we discuss the second and the third conditions. The third condition demands $H = H_{MNP} dx^M \wedge dx^N \wedge dx^P = 0$. Let us recall here the equation we have introduced:

$$H = dB + \omega_{3L} - \omega_{3Y}.$$

By taking the exterior differential of both sides, we obtain,

$$dH = d\omega_{3L} - d\omega_{3Y} = \mathrm{tr}(R \wedge R) - \mathrm{tr}(F \wedge F).$$

Hence the condition $\mathrm{tr}(R \wedge R) = \mathrm{tr}(F \wedge F)$ follows from the condition $H = 0$. What should we do to satisfy the condition $\mathrm{tr}(R \wedge R) = \mathrm{tr}(F \wedge F)$ on K? The simplest answer is to let the $E_8 \times E_8$ gauge field A have the same vacuum expectation value as the spin connection of K. Then the equation $F = R$ holds in the classical limit and the condition $\mathrm{tr}(R \wedge R) = \mathrm{tr}(F \wedge F)$ follows. In this way, we can realize the condition $H = 0$. As for the second condition, it is satisfied by assuming that the holonomy group of K is $SU(3)$ and that $R = F$ holds on K (to be precise, we need discussions on stability of the holomorphic vector bundle, but we omit detailed treatment).

Though the discussions so far are a little bit optimistic, we proceed further by assuming the following two conditions.

- The holonomy group of K is $SU(3)$.
- The gauge field A on K has the same vacuum expectation value as the $SU(3)$ spin connection of K.

The $SU(3)$ vacuum expectation value of the gauge field causes spontaneous symmetry breaking of the gauge group $E_8 \times E_8$. The problem is how this $SU(3)$ is embedded in $E_8 \times E_8$. Here, we assume for simplicity that it is embedded in the left E_8 and that subgroup of E_8 that is commutative with the $SU(3)$ is given by E_6

(this embedding is called diagonal embedding of $SU(3)$ in E_8). From now on, we discuss only the left E_8 in which the $SU(3)$ is embedded. Because of the spontaneous symmetry braking, the gauge group of the 4-dimensional theory reduces to E_6. The group E_6 contains the group $U(1) \times SU(2) \times SU(3)$, which is the gauge group of electroweak and strong interaction, and has been used as a candidate for the gauge group of GUT. Hence our scenario seems promising. Let us consider what kinds of massless fermions appear in this setting. These massless fermions are produced from the fermion field χ^a_α. Since we take note only of the left E_8, subscript a takes a value in adjoint representation **248** of E_8. Under the diagonal embedding, this representation is decomposed into the following representations of $SU(3) \times E_6$.

$$\mathbf{248} = (\mathbf{3}, \mathbf{27}) \oplus (\overline{\mathbf{3}}, \overline{\mathbf{27}}) \oplus (\mathbf{8}, \mathbf{1}) \oplus (\mathbf{1}, \mathbf{78}).$$

In the above formula, **3** is the fundamental representation of $SU(3)$, $\overline{\mathbf{3}}$ is its complex conjugate, **8** is the adjoint representation of $SU(3)$, **78** is the adjoint representation of E_6, **27** is the 27-dimensional representation of E_6 and $\overline{\mathbf{27}}$ is its complex conjugate. We are interested in the $(\mathbf{3}, \mathbf{27}) \oplus (\overline{\mathbf{3}}, \overline{\mathbf{27}})$ part of this decomposition, because in the GUT, which uses E_6 as gauge group, fermions in one generation (a set of two quarks, one lepton and one neutrino) are generated by a fermion in 27-dimensional representation of E_6. On the other hand, if the holonomy group of K is $SU(3)$, K becomes a Kähler manifold, i.e., a complex manifold with Kähler metric $g_{i\bar{j}}$ (the definition of the Kähler manifold will be explained in Chap. 2), and tangent space T_pK at $p \in K$ is decomposed into the direct sum of the holomorphic part and the anti-holomorphic part $T'_pK \oplus \overline{T'_pK}$. In our scenario, $T'K$ and $\overline{T'K}$ correspond to **3** and $\overline{\mathbf{3}}$ representations of $SU(3)$ respectively. In our context, counting the number of 4-dimensional massless fermions which are produced from χ^a_α and take values in $(\mathbf{3}, \mathbf{27})$ representation reduces to counting the dimensions of the Dolbeault cohomology $H^1(T'K)$ of holomorphic tangent bundle $T'K$ (refer to [2] for detailed discussions). Let us explain the outline of the definition of the Dolbeault cohomology $H^q(T'K)$. Let $A_{(0,q)}(T'K)$ be the vector space of $T'K$-valued the differential forms of holomorphic degree 0 and anti-holomorphic degree q. An element of $A_{(0,q)}(T'K)$ is explicitly written as follows:

$$\psi^i_{\bar{j}_1\bar{j}_2\cdots\bar{j}_q} dx^{\bar{j}_1} \wedge dx^{\bar{j}_2} \wedge \cdots \wedge dx^{\bar{j}_q}.$$

Next, we define the anti-holomorphic exterior differential operator $\overline{\partial}_q : A_{(0,q)}(T'K) \to A_{(0,q+1)}(T'K)$ by

$$\overline{\partial}_q(\psi^i_{\bar{j}_1\bar{j}_2\cdots\bar{j}_q} dx^{\bar{j}_1} \wedge dx^{\bar{j}_2} \wedge \cdots \wedge dx^{\bar{j}_q}) := (\partial_{\bar{j}}\psi^i_{\bar{j}_1\bar{j}_2\cdots\bar{j}_q})dx^{\bar{j}} \wedge dx^{\bar{j}_1} \wedge dx^{\bar{j}_2} \wedge \cdots \wedge dx^{\bar{j}_q}.$$

Then the Dolbeault cohomology $H^q(T'K)$ is defined as $\mathrm{Ker}(\overline{\partial}_q)/\mathrm{Im}(\overline{\partial}_{q-1})$. Clearly, $H^q(T'K)$ is trivial if q is greater than 3. For convenience of the following discussion, we introduce the Euler number of holomorphic tangent bundle $T'K$:

$$\chi(T'K) := \sum_{q=0}^{3}(-1)^q \dim(H^q(T'K))$$

Of course, the Dolbeault cohomology $H^1(\overline{T'K})$ can also be considered and its dimension equals the number of 4-dimensional massless fermions which take values in $(\overline{\mathbf{3}}, \overline{\mathbf{27}})$ representation.

Let us explain the properties of the Dolbeault cohomology $H^q(T'K)$. It is known from mathematical results that $H^0(T'K) = H^3(T'K) = 0$. Hence we need information of $H^1(T'K)$ and $H^2(T'K)$. If the holonomy group of K is $SU(3)$, the relation:

$$\dim(H^q(T'K)) = \dim(H^{3-q}((T'K)^*)),$$

follows from the Serre duality theorem which is well-known in complex geometry. In the above relation, $(T'K)^*$ means the dual vector bundle of $T'K$ and it corresponds to the lower subscripts of holomorphic differential forms of K. Since K is a Kähler manifold, a section ψ_i of $(T'K)^*$ is transformed into a section of $\overline{T'K}$ as follows:

$$g^{i\bar{j}}\psi_i = \bar{\psi}^{\bar{j}},$$

where $g^{i\bar{j}}$ is (inverse of) the Kähler metric of K. Hence $(T'K)^*$ is isomorphic to $\overline{T'K}$ as a vector bundle and we have the following relations:

$$\dim(H^1(T'K)) = \dim(H^2(\overline{T'K})), \quad \dim(H^2(T'K)) = \dim(H^1(\overline{T'K})).$$

By combining these results, we obtain

$$\chi(T'K) = -\dim(H^1(T'K)) + \dim(H^2(T'K)) = -\dim(H^1(T'K)) + \dim(H^1(\overline{T'K})).$$

These discussions tell us that the difference between numbers of massless fermions that take values in $(\mathbf{3}, \mathbf{27})$ and $(\overline{\mathbf{3}}, \overline{\mathbf{27}})$ is given by $\chi(T'K)$, which is an invariant of the Kähler manifold K. This fact plays an important role in the discovery of mirror symmetry, and we add some explanation on properties of $\chi(T'K)$. It is known from complex geometry that $\chi(T'K)$ is related to the Euler number $\chi(K)$, which is a topological invariant of K, by the following formula:

$$\chi(T'K) = \frac{1}{2}\chi(K).$$

In the following, we explain how the above relation is derived. $\chi(K)$ is evaluated by the formula $\chi(K) = \sum_{i=0}^{6}(-1)^i \dim(H^i(K))$, where $H^i(K)$ is a vector space of degree i harmonic forms of K. In the case that K is a Kähler manifold, $H^i(K)$ is decomposed as follows:

$$H^i(K) = \bigoplus_{p+q=i} H^{p,q}(K),$$

where $H^{p,q}(K)$ is the vector space of harmonic forms of bi-degree (p,q) (p (resp. q) is holomorphic (resp. anti-holomorphic) degree and $0 \leq p,q \leq 3$ holds). Hence $\chi(K)$ is rewritten as follows:

$$\chi(K) = \sum_{p=0}^{3}\sum_{q=0}^{3}(-1)^{p+q}\dim(H^{p,q}(K)).$$

For simplicity, we introduce the notation $h^{p,q}(K) = \dim(H^{p,q}(K))$. We call the following diagram arranging $h^{p,q}(K)$ in a diamond shape a **Hodge diamond**:

$$\begin{array}{ccccccc} & & & h^{3,3}(K) & & & \\ & & h^{3,2}(K) & & h^{2,3}(K) & & \\ & h^{3,1}(K) & & h^{2,2}(K) & & h^{1,3}(K) & \\ h^{3,0}(K) & & h^{2,1}(K) & & h^{1,2}(K) & & h^{0,3}(K) \\ & h^{2,0}(K) & & h^{1,1}(K) & & h^{0,2}(K) & \\ & & h^{1,0}(K) & & h^{0,1}(K) & & \\ & & & h^{0,0}(K) & & & \end{array}$$

This diagram is convenient for expressing topological characteristics of K as a complex manifold. If K has an $SU(3)$ holonomy group and is a simply connected (this means that any 1-dimensional closed loop in K can continuously shrink into a point) compact manifold, $h^{p,q}(K)$'s are known to satisfy the following relations:

$$h^{p,q}(K) = h^{q,p}(K) = h^{3-p,3-q}(K), \quad h^{0,0}(K) = h^{3,3}(K) = 1,$$
$$h^{1,0}(K) = h^{0,2}(K) = 0.$$

Hence the Hodge diamond of K becomes,

$$\begin{array}{ccccccc} & & & 1 & & & \\ & & 0 & & 0 & & \\ & 0 & & h^{1,1}(K) & & 0 & \\ 1 & & h^{2,1}(K) & & h^{2,1}(K) & & 1\,, \\ & 0 & & h^{1,1}(K) & & 0 & \\ & & 0 & & 0 & & \\ & & & 1 & & & \end{array}$$

and we obtain $\chi(K) = 2(h^{1,1}(K) - h^{2,1}(K))$. Let $\Omega_{ijk}dx^i \wedge dx^j \wedge dx^k$ be a harmonic $(3,0)$-form which is the base of 1-dimensional vector space $H^{3,0}(K)$. It is also a holomorphic differential form called the holomorphic 3-form of K. If we take the representative $\psi^i_{\bar{j}}dx^{\bar{j}}$ of an element of the Dolbeault cohomology $H^1(T'K)$ appropriately, we can make the $(2,1)$-form $\Omega_{ijk}\psi^k_{\bar{l}}dx^i \wedge dx^j \wedge dx^{\bar{l}}$ into a harmonic $(2,1)$-form. It follows that $H^1(T'K)$ is isomorphic to $H^{2,1}(K)$ as a vector space and we obtain $h^{2,1}(K) = \dim(H^1(T'K))$. On the other hand, we have mentioned the relation $H^1(\overline{T'}K) \simeq H^1((T'K)^*)$. But we can further show

$h^{1,1}(K) = \dim(H^1((T'K)^*)) = \dim(H^1(\overline{T'}K))$ by creating a harmonic (1, 1)-form $\phi_{i\bar{j}} dx^i \wedge dx^{\bar{j}}$ from appropriate representative $\phi_{i\bar{j}} dx^{\bar{j}}$ of an element of $H^1((T'K)^*)$. Hence we obtain

$$-\dim(H^1(T'K)) + \dim(H^1(\overline{T'}K)) = -h^{2,1}(K) + h^{1,1}(K) = \frac{1}{2}\chi(K).$$

We have explained that the numbers of 4-dimensional massless fermions in $(\mathbf{3}, \mathbf{27})$ and $(\overline{\mathbf{3}}, \overline{\mathbf{27}})$ are given by $h^{2,1}(K)$ and $h^{1,1}(K)$ respectively. But there exists subtlety here. The difference between these two numbers is interpreted as a kind of anomaly and a $(\mathbf{3}, \mathbf{27})$ massless fermion and a $(\overline{\mathbf{3}}, \overline{\mathbf{27}})$ massless fermion are considered to couple together into a massive fermion. Hence we conclude that the actual number of these massless fermions is given by the absolute value of their difference, i.e., $\frac{1}{2}|\chi(K)|$. As we have mentioned before, one massless fermion of this type corresponds to one generation of fermions in our real world. Therefore, our scenario predicts that the number of generations of fermions is given by $\frac{1}{2}|\chi(K)|$.

1.3 Discovery of Mirror Symmetry of $N = 2$ Superconformal Field Theory

In the previous section, we discussed that the 6-dimensional compact manifold K should have a holonomy group $SU(3)$ in order to obtain a supersymmetric 4-dimensional GUT by compactifying a 10-dimensional $E_8 \times E_8$ heterotic string theory by K. We also showed that the number of generations of fermions in the resulting 4-dimensional theory is given by

$$\frac{1}{2}|\chi(K)| = \frac{1}{2}|\dim(H^1(T'K)) - \dim(H^1(\overline{T'}K))|.$$

In this section, we focus on the manifold K and discuss dynamics of strings moving in K.

First, we discuss the geometrical meaning of the condition that K has a holonomy group $SU(3)$. As was mentioned, it follows from this condition that K is a Käler manifold, i.e., a complex manifold with Kähler metric $g_{i\bar{j}}$. We leave mathematical definitions of the Kähler manifold and Kähler metric to discussion in Chap. 2. Roughly speaking, the condition that K is a Kähler manifold allows us to describe the real 6-dimensional manifold K by three independent complex coordinates (z_1, z_2, z_3) without using their complex conjugates $\overline{z}_1$, $\overline{z}_2$ and $\overline{z}_3$. Moreover, the Kähler metric $g_{i\bar{j}}$ is well-made for constructing an extension of Riemannian geometry in terms of complex coordinates.

To be precise, the condition that K is a Käler manifold is equivalent to the condition that K has a holonomy group $U(3)$. Of course, an element G in $U(3)$ is a 3×3 complex matrix which satisfies the condition ${}^t\overline{G}G = I_3$ (I_3 is a unit matrix of order 3). But we can create a real 6×6 matrix $r(G)$ from G by transforming a complex number $a + b\sqrt{-1}$ into a real 2×2 matrix,

$$\begin{pmatrix} a & -b \\ b & a \end{pmatrix}.$$

The correspondence $G \to r(G)$ gives us an embedding of $U(3)$ in $SO(6)$ (we leave the proof of ${}^t r(G) r(G) = I_6$ and $\det(r(G)) = 1$ to the readers as exercises). From this construction and the fact that $U(3)$ is group of "rotating complex 3-dimensional vectors", we can regard the condition that K has a holonomy group $U(3)$ as the condition that K is a manifold described by 3 complex variables with appropriate metric.

Let us consider an additional restriction on K imposed by the condition that K has a holonomy group $SU(3) \subset U(3)$. In the previous section, we showed that the condition that K has a holonomy group $SU(3)$ is equivalent to the existence of a non–zero spinor η which satisfies the condition $D_i\eta = 0$. Here, D_i is a covariant derivative with respect to the spinor on K. From the condition $D_i\eta = 0$, we can derive

$$[D_i, D_j]\eta = R_{ijkl}\Gamma^{kl}\eta = 0, \tag{1.1}$$

where Γ^{kl} is a Dirac gamma matrix on K and R_{ijkl} is a Riemann curvature tensor. By multiplying (1.1) by gamma matrix Γ^j from the left, we obtain

$$\Gamma^j \Gamma^{kl} R_{ijkl}\eta = 0. \tag{1.2}$$

Then we apply the commutation relation of gamma matrices $\Gamma^j\Gamma^{kl} = \Gamma^{jkl} + g^{jk}\Gamma^l - g^{jl}\Gamma^k$ and the identity of the curvature tensor $R_{ijkl} + R_{iklj} + R_{iljk} = 0$ to (1.2). The result is given by

$$g^{jl}\Gamma^k R_{ijkl}\eta = g^{jl}\Gamma^k R_{jilk}\eta = 0 \Leftrightarrow \Gamma^k R_{ik}\eta = 0. \tag{1.3}$$

Note that $R_{ik} = g^{jl} R_{jilk}$ is the Ricci curvature tensor of K. Since η is not 0 everywhere in K, (1.3) forces the Ricci curvature to vanish everywhere in K. Conversely, it can be shown that K has a holonomy group $SU(3)$ if K has vanishing Ricci curvature. In this way, we obtain the following paraphrase of the condition on K:

- K has a holonomy group $SU(3)$. $\Longleftrightarrow$ K is a complex 3-dimensional Käler manifold with vanishing Ricci curvature.

Now, we proceed to discussion on dynamics of strings moving in K. As was mentioned in the previous section effective low energy theory of $E_8 \times E_8$ heterotic string theory is given by $N = 1$, 10-dimensional supergravity theory coupled with $E_8 \times E_8$ gauge theory. It is difficult to determine the exact theory that describes the motion of a string in K by restricting heterotic string theory to K. But from the point of view of effective theory, we only have to construct a theory of a string moving in K with $N = 1$ supersymmetry. Fundamentally, this theory is constructed by using a map ϕ from Riemann surface Σ to K (explicitly written as $\phi^I(z,\bar{z})$ in terms of local coordinates of Σ and K) and fermion fields $\psi_+^I(z,\bar{z})$, $\psi_-^I(z,\bar{z})$, which are superpartners of $\phi^I(z,\bar{z})$, as dynamical variables. The Lagrangian of this theory is determined by $N=1$ supersymmetry, and it is called the $N=(1,1)$ supersymmetric sigma model on K. $N=(1,1)$ means that there exist a left-handed supercharge Q_+ and a right-handed supercharge Q_- that correspond to z and $\bar{z}$ respectively. Roughly speaking, these supercharges satisfy the following relation:

$$Q_+\phi^I = \psi_+^I, \quad Q_-\phi^I = \psi_-^I.$$

String theorists expected that they could obtain detailed information on the string theory compactified on K by studying this supersymmetric sigma model.

From the discussions on the geometrical condition on K, we know that K is a complex 3-dimensional Käler manifold with vanishing Ricci curvature. This condition adds further symmetries to the supersymmetric sigma model. To get straight to the point, these are given as follows:

- K is a Käler manifold. $\Longrightarrow$ $N=(1,1)$ supersymmetry lifts up to $N=(2,2)$ supersymmetry.
- K has vanishing Ricci curvature. $\Longrightarrow$ Sigma model has conformal invariance.

Let us add some explanations on the first point. $N=(2,2)$ supersymmetry means that the theory has four supercharges $Q_+, Q_-, \overline{Q}_+, \overline{Q}_-$. If K is a Käler manifold, ϕ^I can be written as ϕ^i, $\phi^{\bar{i}}(=\overline{\phi^i})$ by using subscripts of complex coordinates $i=1,2,3$. Accordingly, we can write fermion fields as $\psi_+^i, \psi_-^i, \psi_+^{\bar{i}}, \psi_-^{\bar{i}}$. Then we can operate supercharges on the holomorphic part and on the anti-holomorphic part independently:

$$Q_+\phi^i = \psi_+^i, \quad Q_-\phi^i = \psi_-^i, \quad \overline{Q}_+\phi^{\bar{i}} = \psi_+^{\bar{i}}, \quad \overline{Q}_-\phi^{\bar{i}} = \psi_-^{\bar{i}}.$$

In this way, we can create four supercharges. Of course, original supercharges Q_+ and Q_- are represented as $Q_+ + \overline{Q}_+$ and $Q_- + \overline{Q}_-$ respectively.

Next, we turn to the second point. Conformal invariance means invariance under local scale transformation. In quantum field theory, the response of a theory under local scale transformation is measured by the beta function of renormalization group flow. The beta function of the supersymmetric sigma model on K is shown to be proportional to the Ricci curvature of K. Hence if K has vanishing Ricci curvature, the corresponding sigma model acquires local scale invariance.

At this stage, we introduce some new words for convenience of the following discussions. A Käler manifold with vanishing Ricci curvature is called a Calabi–Yau manifold. On the other hand, a theory on a Riemann surface (real 2-dimensional surface) with supersymmetry and conformal invariance is called a (2-dimensional) superconformal field theory. With these terms, we can summarize discussions so far as follows:

- $N = (1, 1)$ supersymmetric sigma model that describes the motion of a string in complex 3-dimensional Calabi–Yau manifold K becomes an $N = (2, 2)$ superconformal field theory.

In general, an $N = (2, 2)$ superconformal field theory is obtained from gluing a left-handed (holomorphic) $N = 2$ superconformal theory with a right-handed (anti-holomorphic) $N = 2$ superconformal field theory. Therefore, we discuss a left-handed $N = 2$ superconformal theory in the following. The $N = 2$ superconformal field theory has infinite-dimensional symmetry of $N = 2$ superconformal algebra (it is also called $N = 2$ Virasoro algebra), and its structure is highly constrained by this symmetry. Especially in the case of $N = 2$ minimal models, which describe the motion of a string in 0-dimensional spaces, classification of models is completed and their Hilbert spaces are exactly determined. We note here that $N = 2$ superconformal algebra contains $U(1)$ symmetry and the $U(1)$ charge plays an important role in describing the Hilbert space of an $N = 2$ superconformal field theory. In the case of a sigma model on K, the space in which the string moves has positive dimension, and the theory is different from $N = 2$ minimal models. But Gepner, a string theorist, conjectured that an $N = 2$ superconformal theory equivalent to the sigma model on K is obtained from taking the tensor product of several $N = 2$ minimal models. Let us explain the intuition behind Gepner's conjecture. As a tractable example of a complex 3-dimensional Calabi–Yau manifold, we introduce a quintic hypersurface M_5 in 4-dimensional complex projective space CP^4:

$$CP^4 = \{(X_1 : X_2 : X_3 : X_4 : X_5) \mid X_i \in \mathbf{C}\},$$

where the defining equation of the hypersurface is given by

$$(X_1)^5 + (X_2)^5 + (X_3)^5 + (X_4)^5 + (X_5)^5 = 0.$$

On the other hand, a class of $N = 2$ minimal models is classified by a non-negative integer $k = 0, 1, 2, \ldots$, and the theory labeled by k is considered to describe the motion of a string in a complex singularity in $\mathbf{C}^2$ defined by

$$Y^2 = X^{k+2}$$

($(0, 0) \in \mathbf{C}^2$ is the singularity of the curve defined by the above equation). If we set $k = 3$, the defining equation becomes $Y^2 = X^5$. Then we expect that a theory equivalent to the sigma model on M_5 is obtained from taking the tensor product of five copies of the $N = 2$ minimal model labeled by $k = 3$. This conjecture is mysterious from the point of view of mathematics, but it gives us a procedure to determine the Hilbert space of the sigma model on a given Calabi–Yau manifold K. The conjecture was proposed in 1987.

In 1989, two string theorists Greene and Plesser discovered mysterious symmetry when they ware analyzing Hilbert spaces of sigma models on various Calabi–Yau manifolds by using Gepner's conjecture. The outline of this symmetry is explained as follows. We take a complex 3-dimensional Calabi–Yau manifold X and consider a quotient space $X^* = X/G_X$ where G_X is an abelian group determined by X. X^* is also a complex 3-dimensional space with singularities, which is called "orbifold". From orbifold construction of $N = 2$ superconformal field theory, one can compute the Hilbert space of a sigma model on X^*. Let $\mathscr{H}_X$ and $\mathscr{H}_{X^*}$ be Hilbert spaces of sigma models on X and X^* respectively. They found that $\mathscr{H}_X$ and $\mathscr{H}_{X^*}$ are isomorphic as Hibert spaces. Since $N = 2$ superconformal algebra has $U(1)$ symmetry, left- and right-handed $U(1)$-charges (q_L, q_R) are assigned to each state of $N = (2, 2)$ superconformal theory. They also found that the above isomorphism reverses the sign of the right-handed $U(1)$ charge q_R. Let us clarify these assertions by using mathematics terminology. The first assertion is equivalent to the existence of an isomorphism:

$$\varphi : \mathscr{H}_X \xrightarrow{\sim} \mathscr{H}_{X^*}.$$

The second assertion says that if we decompose the Hilbert spaces into

$$\mathscr{H}_X = \bigoplus_{(q_L,q_R)} \mathscr{H}_X^{(q_L,q_R)}, \quad \mathscr{H}_{X^*} = \bigoplus_{(q_L,q_R)} \mathscr{H}_{X^*}^{(q_L,q_R)},$$

the equality:

$$\varphi(\mathscr{H}_X^{(q_L,q_R)}) = \mathscr{H}_{X^*}^{(q_L,-q_R)},$$

holds.

Let us consider what this symmetry implies with respect to geometry of X and X^*. For this purpose, we take note of subspace of the Hilbert space of an $N = (2, 2)$ superconformal theory that consists of chiral primary states. Roughly speaking, chiral primary state means the zero-energy state of an $N = (2, 2)$ superconformal theory. Let $\mathscr{H}_{X,cp}^{(q_L,q_R)}$ and $\mathscr{H}_{X^*,cp}^{(q_L,q_R)}$ be corresponding subspaces of $\mathscr{H}_X^{(q_L,q_R)}$ and $\mathscr{H}_{X^*}^{(q_L,q_R)}$. In the case of a sigma model on a complex 3-dimensional Calabi–Yau manifold, we take the point-particle limit of the string and obtain the decomposition

$$\mathscr{H}_{X,cp}^{(q_L,q_R)} = (\bigoplus_{q_L,q_R=0}^{3} \mathscr{H}_{X,cp,A}^{(q_L,-q_R)}) \oplus (\bigoplus_{q_L,q_R=0}^{3} \mathscr{H}_{X,cp,B}^{(q_L,q_R)}),$$

$$\mathscr{H}_{X^*,cp}^{(q_L,q_R)} = (\bigoplus_{q_L,q_R=0}^{3} \mathscr{H}_{X^*,cp,A}^{(q_L,-q_R)}) \oplus (\bigoplus_{q_L,q_R=0}^{3} \mathscr{H}_{X^*,cp,B}^{(q_L,q_R)}),$$

and the following isomorphisms with Dolbeault cohomolgy,

$$\mathscr{H}_{X,cp,A}^{(q_L,-q_R)} = H^{q_L}(X, \wedge^{q_R}(T'X)^*), \quad \mathscr{H}_{X,cp,B}^{(q_L,q_R)} = H^{q_L}(X, \wedge^{q_R}T'X),$$

$$\mathscr{H}_{X^*,cp,A}^{(q_L,-q_R)} = H^{q_L}(X^*, \wedge^{q_R}(T'X^*)^*), \quad \mathscr{H}_{X^*,cp,B}^{(q_L,q_R)} = H^{q_L}(X^*, \wedge^{q_R}T'X^*).$$

We mentioned above that X^* is an orbifold, but the above Dolbaeult cohomology of X^* means Dolbaeult cohomology of a complex 3-dimensional Calabi–Yau manifold obtained from resolving orbifold singularities of X^*. Since the mysterious symmetry reverses the sign of q_R, it implies the following isomorphism:

$$H^{q_L}(X, \wedge^{q_R}(T'X)^*) \simeq H^{q_L}(X^*, \wedge^{q_R}T'X^*),$$
$$H^{q_L}(X, \wedge^{q_R}T'X) \simeq H^{q_L}(X^*, \wedge^{q_R}(T'X^*)^*). \tag{1.4}$$

By applying the isomorphism derived from Serre duality:

$$H^q(X, \wedge^p T'X) \simeq H^{3-q}(X, \wedge^p(T'X)^*), \quad H^q(X^*, \wedge^p T'X^*) \simeq H^{3-q}(X^*, \wedge^p(T'X^*)^*),$$

we can unify (1.4) into the form:

$$H^q(X^*, \wedge^p(T'X^*)^*) = H^{3-q}(X, \wedge^p(T'X)^*). \tag{1.5}$$

In complex geometry, it is known that the Dolbeault cohomology $H^q(X, \wedge^p(T'X)^*)$ is isomorphic to the vector space of harmonic forms $H^{p,q}(X)$. Hence (1.5) leads us to the following equality:

$$h^{p,q}(X^*) = h^{q,p}(X^*) = h^{q,3-p}(X) = h^{3-p,q}(X), \tag{1.6}$$

where we used the relation $h^{p,q}(X) = h^{q,p}(X)$.

Equation (1.6) suggests interesting properties of topological characteristics of X and X^*. For example, let us compute the Euler numbers of X and X^*.

$$\chi(X^*) = \sum_{p,q=0}^{3} (-1)^{p+q} h^{p,q}(X^*) = \sum_{p,q=0}^{3} (-1)^{p+q} h^{3-p,q}(X) = -\sum_{p,q=0}^{3} (-1)^{p+q} h^{p,q}(X) = -\chi(X).$$

Hence the symmetry reverses the sign of Euler number. Moreover, the Hodge diamond of the complex 3-dimensional Calabi–Yau manifold X is given by

$$\begin{array}{ccccccc} & & & 1 & & & \\ & & 0 & & 0 & & \\ & 0 & & h^{1,1}(X) & & 0 & \\ 1 & & h^{2,1}(X) & & h^{2,1}(X) & & 1 \\ & 0 & & h^{1,1}(X) & & 0 & \\ & & 0 & & 0 & & \\ & & & 1 & & & \end{array}, \tag{1.7}$$

and (1.6) tells us that the Hodge diamond of X^* becomes,

$$\begin{array}{ccccccc} & & & 1 & & & \\ & & 0 & & 0 & & \\ & 0 & & h^{2,1}(X) & & 0 & \\ 1 & & h^{1,1}(X) & & h^{1,1}(X) & & 1 \\ & 0 & & h^{2,1}(X) & & 0 & \\ & & 0 & & 0 & & \\ & & & 1 & & & \end{array}. \tag{1.8}$$

If we compare these two diagrams, we can see that one diagram is obtained from taking a mirror image of the other with respect to a straight line that connects the midpoints of opposite sides of the diamond. Therefore, string theorists began to call the mysterious symmetry "**mirror symmetry**". We say that X^* (resp. X) is mirror manifold of X (resp. X^*), or that X and X^* are mirrors of each other. Let us pick up a concrete example. In the case of the quintic hypersurface M_5 in CP^4, the abelian group G_{M_5} is given as a finite group isomorphic to $(\mathbf{Z}/(5\mathbf{Z}))^3$, which acts on CP^4 as follows:

$$(e^{\frac{2\pi\sqrt{-1}a_1}{5}}X_1 : e^{\frac{2\pi\sqrt{-1}a_2}{5}}X_2 : e^{\frac{2\pi\sqrt{-1}a_3}{5}}X_3 : e^{\frac{2\pi\sqrt{-1}a_4}{5}}X_4 : e^{\frac{2\pi\sqrt{-1}a_5}{5}}X_5),$$

where a_i's are integers that satisfy $\sum_{i=1}^{5} a_i \equiv 0 \pmod 5$. Since this group keeps the defining equation of M_5 invariant, it acts on the hypersurface M_5 and quotient space M_5/G_{M_5} is defined. Since M_5/G_{M_5} is an orbifold, we define M_5^* as a complex 3-dimensional Calabi-Yau manifold obtained from resolving singularities of the orbifold. As for the Euler numbers, we have the results $\chi(M_5) = -200$ and $\chi(M_5^*) = 200$. Therefore, the sign of the Euler number is reversed. We will explain how to compute these numbers in Chap. 2. As for the Hodge numbers, $h^{1,1}(M_5) = h^{2,1}(M_5^*) = 1$ and $h^{2,1}(M_5) = h^{1,1}(M_5^*) = 101$ hold and mirror symmetry of the Hodge diamond is confirmed.

Greene and Plesser constructed several concrete examples of pairs of complex 3-dimensional manifolds X and X^* and proposed mirror symmetry between $N = (2, 2)$ supersymmetric sigma models on X and X^*.

Greene and Plesser suggested an interesting application of mirror symmetry to the supersymmetric sigma model as an $N = 2$ superconformal theory by combining the symmetry with the idea of compactification of heterotic string theory. Let us explain their idea in the following. In the previous section, we mentioned that compactification of heterotic string theory by complex 3-dimensional Calabi–Yau manifold X produces $h^{2,1}(X)$ massless fermions in $(\mathbf{3}, \mathbf{27})$ representation and $h^{1,1}(X)$ massless fermions in $(\overline{\mathbf{3}}, \overline{\mathbf{27}})$ representation. If X^* is a mirror manifold of X, numbers of massless fermions in $(\mathbf{3}, \mathbf{27})$ and $(\overline{\mathbf{3}}, \overline{\mathbf{27}})$ representations are exchanged. One massless fermion in these representations corresponds to one generation of fermions in our 4-dimensional world. Then let us distinguish generations coming from $(\mathbf{3}, \mathbf{27})$ representation by subscripts $\alpha, \beta, \ldots$ and generations from $(\overline{\mathbf{3}}, \overline{\mathbf{27}})$ representation by subscripts $a, b, \ldots$. By the definition of the Hodge number $h^{p,q}(X) = \dim(H^{p,q}(X))$, subscripts $\alpha, \beta, \ldots$ also correspond to the linear independent basis of $H^{2,1}(X)$ and $a, b, \ldots$ correspond to the basis of $H^{1,1}(X)$. In the Lagrangian of 4-dimensional GUT, Yukawa couplings:

$$\lambda_{\alpha\beta\gamma}\phi^{\alpha}\psi^{\beta}\psi^{\gamma}, \quad \lambda_{abc}\phi^{a}\psi^{b}\psi^{c},$$

play important roles. Here, ψ^* (resp. ϕ^*) is a fermion (bosonic) field that belongs to generation $*$ and ϕ^* (existence of bosonic field ϕ^* is assured by 4-dimensional supersymmetry). The coupling constants $\lambda_{\alpha\beta\gamma}$, λ_{abc} are called Yukawa coupling constants (we denote them simply by Yukawa couplings in the following). By detailed analysis of compactification of heterotic string theory, these Yukawa couplings are related to geometrical quantities of the Calabi–Yau manifold used in compactification. Let us first explain the case of $\lambda_{\alpha\beta\gamma}$. We can associate generation α from

$(\mathbf{3}, \mathbf{27})$ representation with a base $u_{(\alpha)}$ of $H^{2,1}(X)$ and by using the isomorphism $H^{2,1}(X) \simeq H^1(T'X)$, we can further associate it with a representative $\tilde{u}^j_{(\alpha),\bar{i}} dx^{\bar{i}}$ of Dolbeault cohomology $H^1(T'X)$. On the other hand, a complex 3-dimensional Calabi–Yau manifold has a holomorphic 3-form $\Omega = \Omega_{ijk} dx^i \wedge dx^j \wedge dx^k$, which corresponds to $h^{3,0}(X) = 1$ and is unique up to multiplication by a constant. It was known among string theorists that $\lambda_{\alpha\beta\gamma}$ can be exactly computed up to multiplication by a constant by the following formula:

$$\lambda_{\alpha\beta\gamma} = \int_X \Omega_{ijk} \tilde{u}^p_{(\alpha),\bar{l}} \tilde{u}^q_{(\beta),\bar{m}} \tilde{u}^r_{(\gamma),\bar{n}} \Omega_{pqr} dx^i \wedge dx^j \wedge dx^k \wedge dx^{\bar{l}} \wedge dx^{\bar{m}} \wedge dx^{\bar{n}}.$$

In the next section, we explain how to use the above formula by taking M_5^* as an example.

Next, we turn to the Yukawa coupling λ_{abc}. Generation a from $(\overline{\mathbf{3}}, \overline{\mathbf{27}})$ representation is associated with a base $v_{(a)}$ of $H^{1,1}(X)$. Of course $v_{(a)}$ is a closed $(1, 1)$-form of X. At the time when Greene and Plesser proposed mirror symmetry, λ_{abc} was known to have the following structure:

$$\lambda_{abc} = \int_X v_{(a)} \wedge v_{(b)} \wedge v_{(c)} + (\text{instanton corrections}).$$

The top term of the r.h.s. can be exactly computed with the aid of complex geometry, but how to determine instanton corrections was not known at that time. The instanton appearing here means world sheet instanton of the sigma model. Mathematically, it is given as a holomorphic map $\phi^i(z)$ from the Riemann sphere (or complex 1-dimensional projective space CP^1) to the Calabi–Yau manifold X. Some algebraic geometers were also interested in this instanton, but it was not known even in algebraic geometry how to compute all the instanton corrections.

At this stage, let us look back at the mirror symmetry. Let $\lambda_{\alpha\beta\gamma}(X)$, $\lambda_{abc}(X)$ be Yukawa couplings obtained from compactification on X and $\lambda_{\alpha\beta\gamma}(X^*)$, $\lambda_{abc}(X^*)$ be the ones obtained from compactification on X^*. Hilbert spaces of sigma models on X and X^* are isomorphic to each other. Hence we can expect that these two theories are isomorphic as $N = 2$ superconformal theories. But generations coming from $(\mathbf{3}, \mathbf{27})$ representation and $(\overline{\mathbf{3}}, \overline{\mathbf{27}})$ are exchanged. These facts lead us to the following heuristic expectation:

- By combining exact results of $\lambda_{\alpha\beta\gamma}(X^*)$ (resp. $\lambda_{\alpha\beta\gamma}(X)$) with the equality expected from the mirror symmetry:

$$\lambda_{\alpha\beta\gamma}(X^*) = \lambda_{abc}(X), \quad \lambda_{\alpha\beta\gamma}(X) = \lambda_{abc}(X^*),$$

can we determine exactly all the instanton corrections of $\lambda_{abc}(X)$ (resp. $\lambda_{abc}(X^*)$)?

In 1991, four string theorists, Candelas, de la Ossa, Green and Parkes carried out this program in the case of $X = M_5$ in their striking paper: "A pair of Calabi–Yau manifolds as an exactly soluble superconformal theory".

1.4 First Striking Prediction of Mirror Symmetry

In this section, we explain the outline of what was done in the paper of Candelas et al. in 1991. As was mentioned in the last section, they took M_5, i.e., quintic hypersurface in CP^4, as the test case of a complex 3-dimensional Calabi–Yau manifold X. We introduced the equation:

$$(X_1)^5 + (X_2)^5 + (X_3)^5 + (X_4)^5 + (X_5)^5 = 0,$$

as the defining equation of M_5, but in general, the defining equation of a quintic hypersurface is given by a degree 5 homogeneous polynomial of X_i $(i = 1, 2, 3, 4, 5)$,

$$F_5 = \sum_{\sum_{i=1}^{5} d_i = 5} a_{d_1 d_2 d_3 d_4, d_5} (X_1)^{d_1} (X_2)^{d_2} (X_3)^{d_3} (X_4)^{d_4} (X_5)^{d_5} = 0.$$

We note here that the coefficients $a_{d_1 d_2 d_3 d_4, d_5}$ should satisfy some subtle conditions imposed from the condition that the quintic hypersurface does not have singularities. If the coefficients $a_{d_1 d_2 d_3 d_4, d_5}$ vary, the shape of the quintic hypersurface as a complex manifold also varies. This variation of shape is directly related to the moduli space of the complex structure of a Calabi–Yau manifold (M_5), which plays a very important role in carrying out the program of mirror symmetry. The moduli space of the complex structure of M_5 is the space which parametrizes the degrees of freedom of giving the complex coordinate system as a complex manifold to M_5. In general, the complex dimension of the moduli space becomes finite and in the case of M_5, it is given by $\dim(H^1(T'M_5)) = \dim(H^{2,1}(M_5)) = h^{2,1}(M_5) = 101$, which follows from the Kodaira-Spencer theory in mathematics. This result seems a little bit hard to understand, but intuitively, the dimension counts the number of substantial parameters of varying the defining equation of M_5. Let us derive the number 101 from this point of view. The number of parameters $a_{d_1 d_2 d_3 d_4, d_5}$ equals the number of degree 5 monomials in X_i $(i = 1, 2, 3, 4, 5)$, which is given as ${}_9C_4 = 126$ by the repeated permutation formula. On the other hand, the linear automorphism of CP^4 given by $X_i \to \sum_{j=1}^{5} b_i^j X_j$ induces variation of the defining equation F_5 but this variation does not change shape of M_5 as a complex manifold. Therefore, $5 \times 5 = 25$ degrees of freedom coming from parameters of linear automorphism is not effective. Hence the number of substantial parameters is given by $126 - 15 = 101$. This computation is heuristic and simple, but if combined with the Kodaira-Spencer theory, it gives us a conventional way to count the Hodge number $h^{2,1}(M_5)$. If X varies in the category of compact complex 3-dimensional Calabi–Yau manifold, only the Hodge numbers $h^{2,1}(X)$ and $h^{1,1}(X)$ vary. In general, the Hodge number $h^{2,1}(X)$ corresponds to the complex dimension of moduli space of the complex structure of X.

In the previous section, we mentioned that for a mirror pair of complex 3-dimensional Calabi–Yau manifolds X and X^*, the relations $h^{2,1}(X) = h^{1,1}(X^*)$ and $h^{1,1}(X) = h^{2,1}(X^*)$ hold. From these relations, we can expect that $h^{1,1}(X)$ also corresponds to the number of parameters related to geometrical characteristics of X.

Roughly speaking, $h^{1,1}(X)$ counts the number of parameters that correspond to "radius" or "size" of X. The corresponding parameter space is called the moduli space of Kähler structure of X. As we have mentioned before, the metric of a Calabi–Yau manifold is given by Kähler metric $g_{i\bar{j}}$. By the definition of a Kähler manifold, the $(1, 1)$-form of X defined by $g_{i\bar{j}}dx^i \wedge dx^{\bar{j}}$ becomes a closed form and the form can be regarded as an element of $H^{1,1}(X)$. Conversely, varying the form in $H^{1,1}(X)$ corresponds to changing the "size" of the Calabi–Yau manifold X.

We have introduced the moduli space of the complex structure of a complex 3-dimensional Calabi–Yau manifold X and the moduli space of the Kähler structure of X and saw that their complex dimensions are given by $h^{2,1}(X)$ and $h^{1,1}(X)$ respectively. By combining these facts with relation between Hodge numbers of X and X^* that are mirrors of each other, we are led to the following conjecture:

- The moduli space of the Kähler structure (resp. complex structure) of X coincides with moduli space of the complex structure (resp. Kähler structure) of X^*.

Based on this conjecture, Candelas et al. tried to carry out the mirror symmetry program in the case of $X = M_5$. First, they took M_5^* as an (a family of) orbifold(s) obtained from dividing the quintic hypersurface defined by

$$F_\psi := (X_1)^5 + (X_2)^5 + (X_3)^5 + (X_4)^5 + (X_5)^5 - 5\psi X_1X_2X_3X_4X_5 = 0,$$

by the abelian group $G_{M_5} \simeq (\mathbf{Z}/(5\mathbf{Z}))^3$ introduced in the previous section. To be precise, we have to resolve singularities of this orbifold but we omit this process because it does not affect the following discussion. Note that the new parameter ψ appears here. It is induced in order to represent degrees of freedom of the moduli space of the complex structure of M_5^* whose dimension is given by $h^{2,1}(M_5^*) = h^{1,1}(M_5) = 1$. It corresponds to a substantial deformation parameter of the defining equation of a quintic hypersurface compatible with action of G_{M_5} and plays the role of a local coordinate of the moduli space. On the other hand, the dimension of the moduli space of the Kähler structure of M_5 also equals 1. Let t be a local coordinate of this moduli space. t is considered to take a complex value and we regularize the parameter t so that the volume of M_5 goes to infinity when we take the limit $\mathrm{Im}(t) \to +\infty$.

At this stage, let us look back at the program in the last part of the previous section. We use t as subscript of the base of $H^{1,1}(M_5)$ and ψ as subscript of the base of $H^{2,1}(M_5^*)$. Since the complex dimensions of the corresponding moduli spaces are both 1, it is feasible to compute all the instanton corrections of the Yukawa coupling $\lambda_{ttt}(M_5)$ by using the exact result of the Yukawa coupling $\lambda_{\psi\psi\psi}(M_5^*)$. What Candelas et al. executed in their striking paper was nothing but this task. From now on, we write the two Yukawa couplings as $\lambda_{ttt}(t)$ and $\lambda_{\psi\psi\psi}(\psi)$ because they turn out to be functions which depend on t and ψ respectively.

As was mentioned in the previous section, the Yukawa coupling $\lambda_{\psi\psi\psi}(\psi)$ can be exactly determined by the formula:

$$\lambda_{\psi\psi\psi}(\psi) = \int_{M_5^*} \Omega_{ijk}\tilde{u}^p_{(\psi),\bar{l}}\tilde{u}^q_{(\psi),\bar{m}}\tilde{u}^r_{(\psi),\bar{n}}\Omega_{pqr}dx^i \wedge dx^j \wedge dx^k \wedge dx^{\bar{l}} \wedge dx^{\bar{m}} \wedge dx^{\bar{n}}. \quad (1.9)$$

Later in this section, we explain the outline of usage of this formula (general discussion will be given in Chap. 4). On the other hand, $\lambda_{ttt}(t)$ was conjectured to have the following expansion form:

$$\lambda_{ttt}(t) = 5 + \sum_{d=1}^{\infty} \alpha_d e^{2\pi i d t}, \tag{1.10}$$

where α_d is the instanton correction coming from degree d world sheet instantons (holomorphic maps) from CP^1 to M_5 and it is given by some real number. If the mirror symmetry program works well, all the α_d's will be determined at once. Let us believe in the conjecture that the moduli space of the Kähler structure of M_5 coincides with the moduli space of the complex structure of M_5^*. Then we can expect the existence of the coordinate transformation:

$$\psi = \psi(t),$$

and the coincidence of the two Yukawa couplings:

$$\lambda_{ttt}(t) = \lambda_{\psi\psi\psi}(\psi(t)) \quad (?!).$$

To be precise, both Yukawa couplings are covariant rank 3 symmetric tensors of the corresponding moduli space and the above equation should be modified into the following form:

$$\lambda_{ttt}(t) = \lambda_{\psi\psi\psi}(\psi(t)) \cdot \left(\frac{d\psi(t)}{dt}\right)^3. \tag{1.11}$$

Fundamentally, they exactly computed $\lambda_{ttt}(t)$ by pushing forward this line of thought. We explain the outline of their computation in the following.

First, we explain the process of computation of $\lambda_{\psi\psi\psi}(\psi)$. In order to use the formula (1.9), we have to compute the holomorphic 3-form $\Omega = \Omega_{ijk} dx^i \wedge dx^j \wedge dx^k$ of M_5^* exactly. Since M_5^* is a quotient space obtained from dividing a quintic hypersurface in CP^4 (defined by $F_\psi = 0$) by an abelian group G_{M_5}, it is given as a holomorphic 3-form of the quintic hypersurface up to multiplication by constant. We take $x_i = \frac{X_i}{X_5}$ $(i = 1, 2, 3, 4)$ as local coordinates of CP^4. A domain that cannot be covered by these local coordinates has measure (volume) 0, and computation can be done by using these local coordinates. The defining equation of the quintic hypersurface is now rewritten as

$$P_\psi := (x_1)^5 + (x_2)^5 + (x_3)^5 + (x_4)^5 + 1 - 5\psi x_1 x_2 x_3 x_4 = 0. \tag{1.12}$$

From the general theory of algebraic geometry, the holomorphic 3-form of the quintic hypersurface is given by

$$-\frac{5\psi}{\frac{\partial P_\psi}{\partial x_4}} dx^1 \wedge dx^2 \wedge dx^3 = \frac{dx^1 \wedge dx^2 \wedge dx^3}{(x_1 x_2 x_3 - \psi^{-1}(x_4)^4)}. \tag{1.13}$$

We note here that x_4 in the above formula is an implicit function in x_1, x_2, x_3 via the equation $P_\psi = 0$. We use this 3-form as the holomorphic 3-form Ω of M_5^*. (1.13) also tells us that Ω depends on the variable ψ. Therefore, we write Ω as $\Omega(\psi)$. In order to proceed to the next stage of computing $\lambda_{\psi\psi\psi}(\psi)$, Candelas et al. considered the period integral $\int_C \Omega(\psi)$ of $\Omega(\psi)$, where C is a homology 3-cycle of M_5^*. Let us remind the readers of the Hodge diamond of M_5^*:

$$
\begin{array}{ccccccc}
 & & & 1 & & & \\
 & & 0 & & 0 & & \\
 & 0 & & 101 & & 0 & \\
1 & & 1 & & 1 & & 1 \\
 & 0 & & 101 & & 0 & \\
 & & 0 & & 0 & & \\
 & & & 1 & & &
\end{array}
$$

From this, we can see that the dimension of $H^3(M_5^*) = \oplus_{p+q=3} H^{p,q}(M_5^*)$ is 4 and that the dimension of homology group $H_3(M_5^*)$ is also 4 (by Poiancaré duality). Hence we can take four homology 3-cycles as the basis of $H_3(M_5^*)$. With the aid of the Hodge theory in complex geometry, they took cycles A^1, A^2, B_1 and B_2 that satisfy the following conditions:

$$A^a \cap B_b = \delta^a_b, \quad A^a \cap A^b = 0, \quad B_a \cap B_b = 0, \tag{1.14}$$

where $C_1 \cap C_2$ means the topological intersection number of 3-cycles C_1 and C_2. For these four 3-cycles, we define the following period integrals:

$$z_1(\psi) = \int_{A^1} \Omega(\psi), \quad z_2(\psi) = \int_{A^2} \Omega(\psi), \quad \mathcal{G}^1(\psi) = \int_{B_1} \Omega(\psi), \quad \mathcal{G}^2(\psi) = \int_{B_2} \Omega(\psi). \tag{1.15}$$

As the basis of the cohomology group $H^3(M_5^*)$, we take α_a and β^a which are Poincaré duals of A^a and B_a respectively. Then we have the relations:

$$\int_{M_5^*} \alpha_a \wedge \beta^b = \delta^b_a, \quad \int_{M_5^*} \alpha_a \wedge \alpha_b = 0, \quad \int_{M_5^*} \beta^a \wedge \beta^b = 0,$$

and we can expand $\Omega(\psi)$ in terms of the basis of $H^3(M_5^*)$ as follows:

$$\Omega(\psi) = z_a(\psi)\beta_a - \mathcal{G}^a(\psi)\alpha_a. \tag{1.16}$$

Since the period integrals are functions in the variable ψ, we can analyze $\Omega(\psi)$ quantitatively by using (1.16). Next, an explicit form of the period integrals is needed. For this purpose, they set the following ansatz of the cycle B_2 by using the local coordinates:

$$B_2 := \{\ |x_1| = |x_2| = |x_3| = \delta,\ x_4 \text{ is determined by the equation } P_\psi = 0, \} \tag{1.17}$$

where δ is a sufficiently small positive real number. By using the technique of residue integrals, they determined the explicit form of the period integral $\mathscr{G}_2(\psi)$:

$$\mathscr{G}_2(\psi) = (\frac{2\pi i}{5})^3 \sum_{n=0}^{\infty} \frac{(5n)!}{(n!)^5 (5\psi)^{5n}}. \tag{1.18}$$

The factor $\frac{1}{5^3}$ is multiplied to represent the operation of dividing by the order of the abelian group G_{M_5}. Let $w_0(\psi)$ be $\sum_{n=0}^{\infty} \frac{(5n)!}{(n!)^5(5\psi)^{5n}}$, which is obtained from eliminating the top factor of $\mathscr{G}_2(\psi)$. It is a solution of the following fourth order linear differential equation:

$$\left(\frac{d^4}{dz^4} - \frac{2(4z-3)}{z(1-z)}\frac{d^3}{dz^3} - \frac{(72z-35)}{5z^2(1-z)}\frac{d^2}{dz^2} - \frac{(24z-5)}{5z^3(1-z)}\frac{d}{dz} - \frac{24}{625z^3(1-z)}\right) w_0(\psi) = 0, \tag{1.19}$$

where $z = \psi^{-5}$. This equation has four linearly independent solutions. It can be shown that the other period integrals are also solutions of the equation. Hence all the period integrals are determined up to the linear combination of the four solutions.

Candelas et al. determined the explicit form of the Yukawa coupling $\lambda_{\psi\psi\psi}(\psi)$ by combining the above differential equation and the following relation:

$$\frac{\partial \Omega(\psi)}{\partial \psi} = \tilde{u}^{p}_{(\psi),\bar{l}} \Omega_{pij}(\psi) dx^i \wedge dx^j \wedge x^{\bar{l}} + f(\psi)\Omega(\psi), \tag{1.20}$$

where $f(\psi)$ is some holomorphic function in ψ. (1.20) is called Kodaira–Spencer equation. The result is given by

$$\lambda_{\psi\psi\psi}(\psi) = C_0 \cdot \frac{\psi^2}{1-\psi^5} \qquad (C_0 \text{ is arbitrary constant}). \tag{1.21}$$

Derivation of this result will be given in Chap. 4.

Let us proceed to the step of finding coordinate transformation $\psi = \psi(t)$. For this purpose, Candelas et al. used information of explicit solutions of (1.19). Equation (1.19) is rewritten in the following form.

$$\left(\left(z\frac{d}{dz}\right)^4 - z\left(z\frac{d}{dz} + \frac{4}{5}\right)\left(z\frac{d}{dz} + \frac{3}{5}\right)\left(z\frac{d}{dz} + \frac{2}{5}\right)\left(z\frac{d}{dz} + \frac{1}{5}\right)\right) w(z) = 0. \tag{1.22}$$

The four linearly independent solutions of (1.22) are given by

$$w_j(z) = \sum_{i=0}^{j} {}_jC_i (\log(z))^j \frac{\partial^{j-i} u(z,0)}{\partial \epsilon^{j-i}},$$
$$\left(u(z,\epsilon) := \sum_{n=0}^{\infty} \frac{\prod_{j=1}^{5n}(j+5\epsilon)}{5^{5n}\prod_{j=1}^{n}(j+\epsilon)^5} z^n\right). \tag{1.23}$$

Derivation of these solutions will also be given in Chap. 4. Note that $\frac{\partial^j u(z,0)}{\partial \epsilon^j}$ is a power series in z and that $w_j(z)$ is written as $\log(z)^j w_0(z) + \cdots$. Since we have used the coordinate $z = \psi^{-5}$ instead of ψ, we have to determine the relation $z = z(t)$, or equivalently $t = t(z)$. At this stage, Candelas et al. considered the location of the point that corresponds to the limit $\mathrm{Im}(t) \to +\infty$, i.e., the point where the volume of M_5 goes to infinity, in the complex z-plane. In conclusion, it turned out to be $z = 0$. Moreover, they imposed the condition that t becomes $t + 1$ if z goes around 0 counterclockwise (to be precise, this condition is a conventional short-cut and their actual discussion is more complicated). This condition is natural because the translation $t \to t + 1$ keeps $\lambda_{ttt}(t)$ invariant, which can be seen from the expansion form in (1.10). When z goes around 0 counterclockwise, the power series in z is invariant, but $\log(z)$ becomes $\log(z) + 2\pi i$. If we try to construct $t(z)$ from the solutions $w_j(z)$, $(j = 0, 1, 2, 3)$, the following function is the most simple choice:

$$t(z) = \frac{1}{2\pi i}\frac{C_1 w_0(z) + w_1(z)}{w_0(z)}$$
$$= \frac{1}{2\pi i}\left(\log(z) + C_1 + \frac{\frac{\partial u(z,0)}{\partial \epsilon}}{w_0(z)}\right), \quad (C_1 \text{ is a constant}). \tag{1.24}$$

The constant C_1 was determined to be $-5\log(5)$ from the global behavior of the solutions. Equation (1.24) and the relation $z = \psi^{-5}$ give us the coordinate transformation $t = t(\psi)$. This transformation is called "mirror map". The inverse transformation $\psi = \psi(t)$ is obtained from the well-known technique of inversion of power series. But it is a little bit complicated and will be explained in Chap. 4. As can be seen from the formula (1.11), the remaining process is to compute $\lambda_{\psi\psi\psi}(\psi(t))\left(\frac{d\psi(t)}{dt}\right)^3$ explicitly. But there exists a subtlety to be considered. As can be seen from the construction of $t = t(\psi)$ in (1.24), period integrals play roles of a kind of homogeneous coordinates of the moduli space of the Kähler structure of M_5. In order to obtain actual coordinates, we have to divide homogeneous coordinates by certain homogeneous coordinate. Equation (1.24) tells us that the homogeneous coordinate used is given by $w_0(\psi)$. Since $\Omega(\psi)$ is linearly expanded in terms of period integrals, we have to divide $\Omega(\psi)$ by $w_0(\psi)$ in translating $\lambda_{\psi\psi\psi}(\psi)$ into $\lambda_{ttt}(t)$. Therefore, we have to divide $\lambda_{\psi\psi\psi}(\psi)$ by $(w_0(\psi))^2$ because it is written as a quadratic form in $\Omega(\psi)$. In this way, they reached the following conclusion:

$$\lambda_{ttt}(t) = \frac{1}{(w_0(\psi(t)))^2}\lambda_{\psi\psi\psi}(\psi(t))\left(\frac{d\psi(t)}{dt}\right)^3$$
$$= \frac{C_0}{(w_0(\psi(t)))^2}\frac{(\psi(t))^2}{1-(\psi(t))^5}\left(\frac{d\psi(t)}{dt}\right)^3. \tag{1.25}$$

Here, the constant C_0 was adjusted to be $5\left(\frac{5}{2\pi i}\right)^3$ so that the constant term of the instanton expansion of $\lambda_{ttt}(t)$ equals 5. By substituting $\psi = \psi(t)$ into (1.25) and expanding the result in powers of $e^{2\pi i t}$, they obtained the following expansion:

$$\lambda_{ttt}(t) = 5 + \sum_{d=1}^{\infty} \alpha_d e^{2\pi i d t} = 5 + 2875 e^{2\pi i t} + 4876875 e^{4\pi i t} + \cdots \tag{1.26}$$

Of course, the instanton correction α_d can be computed as we like by power series calculus. Surprisingly, α_d always turned out to be an integer.

This instanton expansion made a powerful impact on algebraic geometry at that time. The complex 3-dimensional Calabi–Yau manifold M_5 drew the interest of some algebraic geometers from the point of view of enumerative geometry, and Clemens, an algebraic geometer, had proposed the following conjecture:

- The number of rational curves in M_5, which is denoted by n_d, is finite for arbitrary positive integer d.

Roughly speaking, a rational curve of degree d in M_5 is a complex 1-dimensional submanifold of M_5 of degree d, which is given as the image of a holomorphic map from CP^1 to M_5. Following this conjecture, n_1 and n_2 were explicitly computed at that time. They are given by

$$n_1 = 2875, \quad n_2 = 609250. \tag{1.27}$$

n_d for higher d had not been computed because the process of computation by algebraic geometry got extremely complicated as the degree d got higher. If we compare (1.26) with (1.27), we immediately notice $n_1 = \alpha_1$. Then how about n_2? α_2 and n_2 are different from each other, but the following equality holds:

$$4876875 = 2^3 \cdot 609250 + 2875.$$

Inspired by this equality, they proposed the following conjecture:

n_d determined from the equality

$$\lambda_{ttt}(t) = 5 + \sum_{d=1}^{\infty} \alpha_d e^{2\pi i d t} = 5 + \sum_{d=1}^{\infty} \frac{n_d d^3 e^{2\pi i d t}}{1 - e^{2\pi i d t}}, \tag{1.28}$$

is nothing but the number of degree d rational curves in M_5.

In fact, n_d obtained from (1.28) always turned out to be a positive integer. This condition is more non-trivial than the condition that α_d is a positive integer. Soon after their paper appeared, the conjecture for the $d = 3$ case was confirmed by algebraic geometers. This conjecture was striking to algebraic geometers, because they did not use any technique of enumerative geometry but used mysterious "mirror symmetry", which was discovered from pursuing compactification of heterotic string theory by Calabi–Yau manifolds.

References

1. P. Candelas, X.C. de la Ossa, P.S. Green, L. Parkes, *A pair of Calabi-Yau manifolds as an exactly soluble superconformal theory.* Nuclear Phys. B 359 (1991), no. 1, 21–74
2. M.B. Green, J.H. Schwarz, E. Witten. *Superstring Theory, Volume 2, Loop Amplitudes, Anomalies and Phenomenology.* Cambridge Monographs on Mathematical Physics, Cambridge University Press (1988)

Chapter 2
Basics of Geometry of Complex Manifolds

Abstract In this chapter, we explain the basics of geometry of complex manifolds, which will help the readers to understand the results of mirror symmetry. First, we introduce the definition of complex manifolds and holomorphic vector bundles on complex manifolds. We also discuss Chern classes of holomorphic vector bundles. Then we introduce Kähler manifolds, which play a central role in geometry of complex manifolds, and explain various characteristics of projective space, which is the most important example in this book. Lastly, we discuss the outline of the notion of an orbifold, which appears in various aspects of mirror symmetry. (For deeper understanding of complex geometry, we recommend readers to consult [1].)

2.1 Complex Manifold

Let us introduce the definition of a complex manifold M.

Definition 2.1.1 A topological space M is a complex manifold if M satisfies the following three conditions:
(i) M is a Hausdorff space.
(ii) M has a countable open covering $M = \bigcup_{\alpha} U_\alpha$ and for each U_α, there exists a homeomorphism $\phi_\alpha : U_\alpha \to \mathscr{U}_\alpha$, which maps U_α to open subset $\mathscr{U}_\alpha$ of $\mathbf{C}^n$.
(iii) If $U_\alpha \cap U_\beta \neq \emptyset$, $\phi_\beta \circ \phi_\alpha^{-1} : \phi_\alpha(U_\alpha \cap U_\beta) \to \phi_\beta(U_\alpha \cap U_\beta)$ is a biholomorphic map.

Let us discuss the difference between the standard definition of a C^∞ $2n$-dimensional real manifold and the above definition. In the condition (ii), we use $\mathbf{C}^n$ instead of R^{2n}, i.e.,

$$\phi_\alpha(p) = (z_\alpha^1(p), z_\alpha^2(p), \cdots, z_\alpha^n(p)), \quad z_\alpha^j(p) \in \mathbf{C}, \quad (j = 1, 2, \cdots, n). \quad (2.1)$$

If we write $z_\alpha^j(p) = x_\alpha^j(p) + i y_\alpha^j(p)$ and regard $z_\alpha^j(p)$ as a pair of two real coordinates $(x_\alpha^j(p), y_\alpha^j(p))$, we can consider M as an example of a C^∞ real $2n$-dimensional

M. Jinzenji, *Classical Mirror Symmetry*, SpringerBriefs in Mathematical Physics,
https://doi.org/10.1007/978-981-13-0056-1_2

manifold. But the crucial difference is given by the condition (iii). This condition says that the coordinate transformation

$$z^j_\beta = z^j_\beta(z^1_\alpha, \cdots, z^n_\alpha), \quad (j = 1, 2, \cdots, n), \tag{2.2}$$

and its inverse

$$z^j_\alpha = z^j_\alpha(z^1_\beta, \cdots, z^n_\beta), \quad (j = 1, 2, \cdots, n), \tag{2.3}$$

are both holomorphic in $(z^1_\alpha, \cdots, z^n_\alpha)$ and $(z^1_\beta, \cdots, z^n_\beta)$. Let us explain the intuitive image of a holomorphic map. We set $n = 1$ for simplicity and consider a coordinate transformation

$$x_\beta = x_\beta(x_\alpha, y_\alpha), \ y_\beta = y_\beta(x_\alpha, y_\alpha), \tag{2.4}$$

in two real variables. If we set $x_\beta + iy_\beta = z_\beta$, we can rewrite the above transformation into the form:

$$z_\beta = z_\beta(x_\alpha, y_\alpha). \tag{2.5}$$

If we try to express x_α and y_α in terms of $z_\alpha = x_\alpha + iy_\alpha$, we are forced to introduce the complex conjugate variable $\overline{z}_\alpha = x_\alpha - iy_\alpha$ because x_α and y_α are independent. Then we can rewrite the above transformation into the form:

$$z_\beta = z_\beta\left(\frac{z_\alpha + \overline{z}_\alpha}{2}, \frac{z_\alpha - \overline{z}_\alpha}{2i}\right) = z_\beta(z_\alpha, \overline{z}_\alpha). \tag{2.6}$$

From this operation, we can see that **there exists a possibility of complex conjugate variable $\overline{z}_\alpha$ appearing if we try to express a general C^∞ 2-dimensional coordinate transformation in terms of a complex variable.** A holomorphic map is a map that is independent of a complex conjugate variable

$$z_\beta = z_\beta(z_\alpha), \tag{2.7}$$

such as $z_\beta = z_\alpha + (z_\alpha)^2$. In the case of multi-variables like (2.3), it means that the r.h.s. does not depend on $\overline{z}^j_\alpha, \quad (j = 1, \ldots, n)$. In the $n = 1$ case, it is obvious that the following coordinate transformation is not holomorophic:

$$x_\beta = x_\alpha, \ y_\beta = -y_\alpha. \tag{2.8}$$

A complex manifold must satisfy the condition that **each local complex coordinate system should be glued by a holomorphic coordinate transformation.**

Let us consider the case of real 2-dimensional closed manifolds. We can easily see that the holomorphic coordinate transformation preserves the orientation of R^2. Therefore, an unorientable closed surface like a Klein bottle cannot be a complex manifold. On the other hand, it is proved that every orientable closed surface can have a complex coordinate system of a 1-dimensional complex manifold. This example is the so-called Riemann surface. But the statement that every orientable $2n$-dimensional closed manifold can be a complex manifold is not true. For example, 4-dimensional sphere S^4 cannot be a complex manifold.

Let us introduce an n-dimensional projective space CP^n as an important example of a compact complex manifold. CP^n is a manifold obtained from the set:

$$(\mathbf{C}^{n+1} - \{\mathbf{0}\}) = \{(X_1, X_2, \ldots, X_n, X_{n+1}) \in \mathbf{C}^{n+1} \mid (X_1, X_2, \ldots, X_{n+1}) \neq (0, 0, \ldots, 0)\},$$

by imposing the following equivalence relation:

$$(X_1, X_2, \ldots, X_n, X_{n+1}) \simeq (\lambda X_1, \lambda X_2, \ldots, \lambda X_n, \lambda X_{n+1}), \quad (\lambda \in \mathbf{C}^{\times}). \quad (2.9)$$

This equivalence relation corresponds to taking the ratio of homogeneous coordinates X_i's, and we alternatively represent CP^n by

$$CP^n = \{\ (X_1 : X_2 : \cdots : X_{n+1})\ \}. \quad (2.10)$$

(Here, $(0 : 0 : \cdots : 0)$ is excluded.) If we take open subsets $U_{(i)} := \{\ (X_1; X_2 : \cdots : X_{n+1}) \in CP^n \mid X_i \neq 0\ \}$, we obtain an open covering $\bigcup_{i=1}^{n+1} U_{(i)} = CP^n$, and we can construct the following complex coordinate system $\phi_{(i)} : U_{(i)} \to \mathbf{C}^n$:

$$\phi_{(i)}(X_1 : X_2 : \cdots : X_{n+1}) = \left(\frac{X_1}{X_i}, \ldots, \frac{X_{i-1}}{X_i}, \frac{X_{i+1}}{X_i}, \ldots, \frac{X_{n+1}}{X_i}\right) = (z^1_{(i)}, z^2_{(i)}, \ldots, z^n_{(i)}). \quad (2.11)$$

In this set-up, we can easily see that $\phi_{(j)} \circ \phi^{-1}_{(i)}$ is biholomorphic in $U_{(i)} \cap U_{(j)}$. For example, $\phi_{(2)} \circ \phi^{-1}_{(1)}$is given by

$$z^1_{(2)} = \frac{1}{z^1_{(1)}}, \quad z^j_{(2)} = \frac{z^j_{(1)}}{z^1_{(1)}}, \quad (j = 2, 3, \cdots n). \quad (2.12)$$

The above map and its inverse are holomorphic since $z^1_1 = \frac{X_2}{X_1} \neq 0$ in $U_{(1)} \cap U_{(2)}$. For general i, j, $\phi_{(j)} \circ \phi^{-1}_{(i)}$ is given by

$$(i < j)\ \ z^m_{(j)} = \frac{z^m_{(i)}}{z^{j-1}_{(i)}}, \ (1 \leq m \leq i-1), \ \ z^m_{(j)} = \frac{1}{z^{j-1}_{(i)}}, \ (m = i),$$

$$z^m_{(j)} = \frac{z^{m-1}_{(i)}}{z^{j-1}_{(i)}}, \ (i+1 \leq m \leq j-1), \ \ z^m_{(j)} = \frac{z^m_{(i)}}{z^{j-1}_{(i)}}, \ (j \leq m \leq n). \quad (2.13)$$

$$(j < i)\ z_{(j)}^m = \frac{z_{(i)}^m}{z_{(i)}^j},\ (1 \leq m \leq j-1),\ z_{(j)}^m = \frac{z_{(i)}^{m+1}}{z_{(i)}^j},\ (j \leq m \leq i-2),$$

$$z_{(j)}^m = \frac{1}{z_{(i)}^j},\ (m = i-1),\ z_{(j)}^m = \frac{z_{(i)}^m}{z_{(i)}^j},\ (i \leq m \leq n). \tag{2.14}$$

As the next example, we pick up a degree k hypersurface in CP^n. Let us consider a degree k homogeneous polynomial in $X_1, \ldots, X_{n+1}$:

$$F(X_1, \ldots, X_{n+1}) = \sum_{\sum_{j=1}^{n+1} d_j = k} a_{(d_1,\ldots,d_{n+1})} \prod_{j=1}^{n+1} (X_j)^{d_j}. \tag{2.15}$$

Here, we impose the condition that the following simultaneous equation

$$\frac{\partial F}{\partial X_i}(X_1, \ldots, X_{n+1}) = 0,\ \ (i = 1, 2, \ldots, n+1), \tag{2.16}$$

does not have solutions except for $(X_1, \ldots, X_{n+1}) = (0, 0, \ldots, 0)$. Since we have

$$F(\lambda X_1, \ldots, \lambda X_{n+1}) = (\lambda)^k F(X_1, \ldots, X_{n+1}), \tag{2.17}$$

the equation $F(X_1, X_2, \ldots, X_{n+1}) = 0$ imposes a constraint that does not contradict (2.9). Then the degree k hypersurface in CP^n is defined by

$$\{(X_1; X_2 : \cdots : X_{n+1}) \in CP^n \mid F(X_1, X_2, \ldots, X_{n+1}) = 0\}. \tag{2.18}$$

The condition with respect to (2.16) assures us that the hypersurface does not have singularities (points whose open neighborhood is not biholomorphic to an open neighborhood of $\mathbf{C}^{n-1}$). Under this condition, the degree k hypersurface becomes an $(n-1)$-dimensional complex manifold. Let us denote the degree k hypersurface in CP^n by M_n^k.

2.2 Vector Bundles of Complex Manifold

Let us assume that all the maps appearing in this section are C^∞ maps.

2.2.1 *Definition of Holomorphic Vector Bundles, Covariant Derivatives, Connections and Curvatures*

Let us introduce a complex vector bundle E on a complex manifold M.

Definition 2.2.1 A complex manifold E is a rank r complex vector bundle on an n-dimensional complex manifold M if it satisfies the following conditions:
(i) There exists a surjection $\pi : E \to M$.
(ii) For any $x \in M$, $\pi^{-1}(x)$ is a complex vector space linearly isomorphic to $\mathbf{C}^r$.
(iii) There exists an open covering $M = \bigcup_{\alpha} U_\alpha$ and for each U_α, we have a homeomorphism $\phi_\alpha : \pi^{-1}(U_\alpha) \to U_\alpha \times \mathbf{C}^r$. Moreover, for any $x \in U_\alpha$, $\phi_\alpha|_{\pi^{-1}(x)} : \pi^{-1}(x) \to \mathbf{C}^r$ is a linear isomorphism between complex vector spaces.

Let us illustrate an intuitive image of a complex vector bundle E on M. First, let us consider a vector-valued function f on M: $f : M \to \mathbf{C}^r$. As you know from high school mathematics, it is convenient to introduce a graph of f to grasp an intuitive image of the function f. In modern language, a graph of f is given by a map $graph_f : M \to M \times \mathbf{C}^r$ where $graph_f(x) = (x, f(x))$. A rank r complex vector bundle E is a generalization of the target space of $graph(f)$: $M \times \mathbf{C}^r$. Roughly speaking, if we restrict our attention to an open subset U_α of M, E can be regarded as a target space of $graph_f$: $U_\alpha \times \mathbf{C}^r$. But in the case of a complex vector bundle E, we add a degree of freedom to "twist" $\mathbf{C}^r$ globally along M. To illustrate with an example, let us consider a rank 1 complex vector bundle E on CP^1. As we have seen in the previous section, CP^1 is covered by two open subsets $U_{(0)} = \{(X_0 : X_1) \mid X_0 \neq 0\}$ and $U_{(1)} = \{(X_0 : X_1) \mid X_1 \neq 0\}$. The complex coordinate system of CP^1 is given by,

$$\phi_{(0)}(X_0 : X_1) = \frac{X_1}{X_0} = z, \quad \phi_{(1)}(X_0 : X_1) = \frac{X_0}{X_1} = w. \tag{2.19}$$

In this coordinate system, $U_{(0)} \cap U_{(1)}$ is isomorphic to $\mathbf{C}^\times$ given by $\{z \in \mathbf{C} | z \neq 0\}$ or $\{w \in \mathbf{C} | w \neq 0\}$. Following Definition 2.2.1, we introduce

$$\phi_0 : \pi^{-1}(U_{(0)}) \to \mathbf{C} \times \mathbf{C}, \quad \phi_1 : \pi^{-1}(U_{(1)}) \to \mathbf{C} \times \mathbf{C}. \tag{2.20}$$

where we have identified $U_{(0)}$ and $U_{(1)}$ with $\mathbf{C} = \{z\}$ and $\mathbf{C} = \{w\}$ respectively. Then let us consider $\phi_1 \circ \phi_0^{-1} : \mathbf{C}^\times \times \mathbf{C} \to \mathbf{C}^\times \times \mathbf{C}$, where $C^\times$ is identified with $U_{(0)} \cap U_{(1)}$. If we define $g_{10}(z) \in \mathbf{C}^\times$ by $\phi_1 \circ \phi_0^{-1}(z, v) = g_{10}(z)v$, g_{10} can be regarded as a map from $\mathbf{C}^\times$ to $\mathbf{C}^\times$. This is the so-called transition function of E. If we set $g_{10}(z) = z^0 = 1$, this operation corresponds to no global twisting. Then E is identified with $CP^1 \times \mathbf{C}$. But what occurs if we set $g_{10}(z) = z^m$ with non-zero integer m? We have to note that it is well-defined even for negative m because z never vanishes in $U_{(0)} \cap U_{(1)}$. As a map from $\mathbf{C}^\times$ to $\mathbf{C}^\times$, z^m and z^n are topologically different (not homotopic) if $m \neq n$. Therefore, global twisting occurs in this case and E is topologically distinguishable from $CP^1 \times \mathbf{C}$. This is an example of a non-trivial complex vector bundle.

Let us get back to the map $graph_f : M \to M \times \mathbf{C}^r$ associated with the vector-valued function $f : M \to \mathbf{C}^r$. If we introduce a projection $\pi : M \times \mathbf{C}^r \to \mathbf{C}^r$, we can reconstruct f from $graph_f$ via the equality $f = \pi \circ graph_f$. Therefore, we can use the map $graph_f : M \to M \times \mathbf{C}^r$ to represent the vector valued function f.

With this fact in mind, we introduce the following definition as a generalization of a vector-valued function to the vector bundle E on M.

Definition 2.2.2 For a complex vector bundle E on a complex manifold M, a section of E is a map $f : M \to E$ which satisfies $\pi \circ f = id_M$.

So far, we have explained that a rank r complex vector bundle E on M is a geometrical object obtained from twisting $\mathbf{C}^r$ of $M \times \mathbf{C}^r$ globally along M. Let us introduce a function which represents how $\mathbf{C}^r$ is twisted globally along M. The map $\phi_\alpha : \pi^{-1}(U_\alpha) \to U_\alpha \times \mathbf{C}^r$ that appeared in Definition 2.2.1 is called a local trivialization of E on U_α. If $U_\alpha \cap U_\beta \neq \emptyset$, we can consider a map $\phi_\beta \circ \phi_\alpha^{-1} : U_\alpha \cap U_\beta \times \mathbf{C}^r \to U_\alpha \cap U_\beta \times \mathbf{C}^r$. We can regard its restriction $\phi_\beta \circ \phi_\alpha^{-1}|_{\{x\}\times\mathbf{C}^r}$ as a linear isomorphism from $\{x\} \times \mathbf{C}^r$ to $\{x\} \times \mathbf{C}^r$. Therefore, we can represent it by using $g_{\beta\alpha}(x) \in GL(r, \mathbf{C})$:

$$\phi_\beta \circ \phi_\alpha^{-1}(x, e_a^{(\alpha)}) = (x, e_b^{(\beta)}(g_{\beta\alpha}(x))^b_a), \tag{2.21}$$

where $e_a^{(\alpha)}$'s and $e_b^{(\beta)}$'s are the standard basis of $U_\alpha \times \mathbf{C}^r$ and $U_\beta \times \mathbf{C}^r$ respectively. From now on, we abbreviate the above equality as follows:

$$e_a^{(\alpha)} = e_b^{(\beta)}(g_{\beta\alpha}(x))^b_a. \tag{2.22}$$

A section $f : M \to E$ is represented as a vector-valued function $(A^1_{(\alpha)}(x), \ldots, A^r_{(\alpha)}(x))$ on U_α by using $\phi_\alpha \circ f(x) = (x, (A^1_{(\alpha)}(x), \ldots, A^r_{(\alpha)}(x)))$. Since $e_a^{(\alpha)} A^a_{(\alpha)}(x) = e_b^{(\beta)} A^b_{(\beta)}(x)$ must hold on $U_\alpha \cap U_\beta$, we obtain

$$e_a^{(\alpha)} A^a_{(\alpha)}(x) = e_b^{(\beta)}(g_{\beta\alpha}(x))^b_a A^a_{(\alpha)}(x) = e_b^{(\beta)} A^b_{(\beta)}(x) \Longleftrightarrow (g_{\beta\alpha}(x))^b_a A^a_{(\alpha)}(x) = A^b_{(\beta)}(x). \tag{2.23}$$

The map $g_{\beta\alpha} : U_\alpha \cap U_\beta \to GL(r, \mathbf{C})$ quantitatively represents how E is twisted globally along M and is called a "**transition function**" of E. In the case of a trivial vector bundle $M \times \mathbf{C}^r$, we can make $g_{\beta\alpha}(x)$ unit matrix I_r for any $x \in U_\alpha \cap U_\beta$ and any $U_\alpha \cap U_\beta \neq \emptyset$ by taking $\{e_a^{(\alpha)}\}$ and $\{e_b^{(\beta)}\}$ as the global standard base of $\mathbf{C}^r$. We can also see from (2.23) that the transition function plays the role of translating the local representation of the section $e_a^{(\alpha)} A^a_{(\alpha)}(x)$ on U_α into that of U_β. At this stage, we can introduce the notion of holomorphic vector bundles.

Definition 2.2.3 Let M be an n-dimensional complex manifold M and E be a rank r complex vector bundle on M. We assume that the open covering $\cup_\alpha U_\alpha = M$ allows both the local holomorphic coordinate system of M and local trivialization of E. A rank r holomorphic vector bundle E on M is a holomorphic vector bundle when we can take a local trivialization of E such that all the transition functions $g_{\alpha\beta} : U_\alpha \cap U_\beta \to GL(r, \mathbf{C})$ are holomorphic.

Next, we introduce the covariant derivative ∇_i, $\nabla_{\bar{i}}$ of a section f of holomorphic vector bundle E. As we have discussed before, f is represented by $e_a^{(\alpha)} A^a_{(\alpha)}$ on U_α.

Then we take holomorphic coordinates $(z^1_{(\alpha)}, z^2_{(\alpha)}, \ldots, z^n_{(\alpha)})$ on U_α and define $\nabla^{(\alpha)}_i$ and $\nabla^{(\alpha)}_{\bar{i}}$ as follows:

$$\begin{aligned}
\nabla^{(\alpha)}_i (e^{(\alpha)}_a A^a_{(\alpha)}) &:= (\nabla^{(\alpha)}_i e^{(\alpha)}_a) A^a_{(\alpha)} + e^{(\alpha)}_a (\partial_i A^a_{(\alpha)}), \\
\nabla^{(\alpha)}_{\bar{i}} (e^{(\alpha)}_a A^a_{(\alpha)}) &:= (\nabla^{(\alpha)}_{\bar{i}} e^{(\alpha)}_a) A^a_{(\alpha)} + e^{(\alpha)}_a (\partial_{\bar{i}} A^a_{(\alpha)}),
\end{aligned} \tag{2.24}$$

where we represent $\dfrac{\partial A^a_{(\alpha)}}{\partial z^i_{(\alpha)}}$ and $\dfrac{\partial A^a_{(\alpha)}}{\partial \overline{z^i_{(\alpha)}}}$ by $\partial_i A^a_{(\alpha)}$ and $\partial_{\bar{i}} A^a_{(\alpha)}$ respectively. We have to note that $\nabla^{(\alpha)}_i e^{(\alpha)}_a$ and $\nabla^{(\alpha)}_{\bar{i}} e^{(\alpha)}_a$ appear in (2.24). This is because $\{e^{(\alpha)}_a\}$ is a local basis used for local trivialization and not a globally universal basis. In the case of holomorphic vector bundle E with holomorphic local trivialization, we can make $\nabla^{(\alpha)}_{\bar{i}} e^{(\alpha)}_a$ vanish, but in general, $\nabla^{(\alpha)}_i e^{(\alpha)}_a$ remains non-zero. Therefore, we set these quantities as follows:

$$\nabla^{(\alpha)}_i e^{(\alpha)}_a = e^{(\alpha)}_b (\omega^{(\alpha)}_i)^b_a, \quad \nabla^{(\alpha)}_{\bar{i}} e^{(\alpha)}_a = 0. \tag{2.25}$$

We call the matrix-valued function $\omega^{(\alpha)}_i$ a **connection** of holomorphic vector bundle E on U_α. We can determine the relation between $\omega^{(\alpha)}_i$ and $\omega^{(\alpha)}_i$ $(U_\alpha \cap U_\beta = \emptyset)$ by imposing $\nabla^{(\alpha)}_i e^{(\alpha)}_a = \nabla^{(\beta)}_i e^{(\alpha)}_a$. In the following computation, we use holomorphic coordinates on U_α to avoid needless complexity.:

$$\begin{aligned}
&\nabla^{(\alpha)}_i e^{(\alpha)}_a = \nabla^{(\beta)}_i e^{(\alpha)}_a = \nabla^{(\beta)}_i (e^{(\beta)}_b (g_{\beta\alpha})^b_a) = (\nabla^{(\beta)}_i e^{(\beta)}_b)(g_{\beta\alpha})^b_a + e^{(\beta)}_b (\partial_i g_{\beta\alpha})^b_a \\
&= e^{(\beta)}_b (\omega^{(\beta)}_i)^b_c (g_{\beta\alpha})^c_a + e^{(\beta)}_b (\partial_i g_{\beta\alpha})^b_a = e^{(\alpha)}_c (\omega^{(\alpha)}_i)^c_a = e^{(\beta)}_b (g_{\beta\alpha})^b_c (\omega^{(\alpha)}_i)^c_a, \\
&\Longleftrightarrow (\omega^{(\beta)}_i)^b_c (g_{\beta\alpha})^c_a + (\partial_i g_{\beta\alpha})^b_a = (g_{\beta\alpha})^b_c (\omega^{(\alpha)}_i)^c_a, \\
&\Longleftrightarrow (g^{-1}_{\beta\alpha})^d_b (\omega^{(\beta)}_i)^b_c (g_{\beta\alpha})^c_a + (g^{-1}_{\beta\alpha})^d_b (\partial_i g_{\beta\alpha})^b_a = (\omega^{(\alpha)}_i)^d_a.
\end{aligned} \tag{2.26}$$

In matrix form, the above relation can be written as follows:

$$(\omega^{(\alpha)}_i) = (g^{-1}_{\beta\alpha})(\omega^{(\beta)}_i)(g_{\beta\alpha}) + (g^{-1}_{\beta\alpha})(\partial_i g_{\beta\alpha}). \tag{2.27}$$

As for $\omega^{(\alpha)}_{\bar{i}}$, we can set $\omega^{(\alpha)}_{\bar{i}} = 0$ because we have $(g^{-1}_{\beta\alpha})(\partial_{\bar{i}} g_{\beta\alpha}) = 0$.

To prepare for future applications, we introduce a positive Hermitian metric on E and discuss a connection which is compatible with the metric. We first introduce the Hermitian metric $h^{(\alpha)}_{a\bar{b}}$ on $E|_{U_\alpha}$ by considering the Hermitian inner product of two sections $e^{(\alpha)}_a A^a_{(\alpha)}$ and $e^{(\alpha)}_b B^b_{(\alpha)}$ on U_α:

$$\langle e^{(\alpha)}_a A^a_{(\alpha)}, e^{(\alpha)}_b B^b_{(\alpha)} \rangle = A^a_{(\alpha)} \overline{B^b_{(\alpha)}} \langle e^{(\alpha)}_a, e^{(\alpha)}_b \rangle = A^a_{(\alpha)} \overline{B^b_{(\alpha)}} h^{(\alpha)}_{a\bar{b}}. \tag{2.28}$$

In (2.28), we use the subscript $\bar{b}$ in $h^{(\alpha)}_{a\bar{b}}$ because we take the complex conjugate in the right entry of the Hermitian metric. Note that $h^{(\alpha)}_{a\bar{b}} = \overline{h^{(\alpha)}_{b\bar{a}}}$ holds since we have the relation $\langle e^{(\alpha)}_a, e^{(\alpha)}_b \rangle = \overline{\langle e^{(\alpha)}_b, e^{(\alpha)}_a \rangle}$. From the global compatibility of the Hermitian matric, we can derive the relation between $h^{(\alpha)}_{a\bar{b}}$ and $h^{(\beta)}_{a\bar{b}}$ in $U_\alpha \cap U_\beta = \emptyset$:

$$h^{(\alpha)}_{a\bar{b}} = \langle e^{(\alpha)}_a, e^{(\alpha)}_b \rangle = \langle e^{(\beta)}_c (g_{\beta\alpha})^c_a, e^{(\beta)}_d (g_{\beta\alpha})^d_b \rangle = \langle e^{(\beta)}_c, e^{(\beta)}_d \rangle (g_{\beta\alpha})^c_a \overline{(g_{\beta\alpha})^d_b} = h^{(\beta)}_{c\bar{d}} (g_{\beta\alpha})^c_a \overline{(g_{\beta\alpha})^d_b}. \tag{2.29}$$

At this stage, we introduce the contravariant metric $h^{a\bar{b}}_{(\alpha)}$ defined as follows:

$$h^{a\bar{b}}_{(\alpha)} h^{(\alpha)}_{c\bar{b}} = \delta^c_a, \quad h^{a\bar{b}}_{(\alpha)} h^{(\alpha)}_{a\bar{c}} = \delta^{\bar{b}}_{\bar{c}}. \tag{2.30}$$

Then we define a connection of E which is compatible with the Hermitian metric. The compatibility condition is given by

$$\partial_i h^{(\alpha)}_{a\bar{b}} = \langle (\nabla^{(\alpha)}_i e^{(\alpha)}_a), e^{(\alpha)}_b \rangle = \langle e^{(\alpha)}_c (\omega^{(\alpha)}_i)^c_a, e^{(\alpha)}_b \rangle = (\omega^{(\alpha)}_i)^c_a h^{(\alpha)}_{c\bar{b}}. \tag{2.31}$$

In (2.31), we have to note that $\nabla^{(\alpha)}_{\bar{i}} e^{(\alpha)}_b = 0 \Leftrightarrow \nabla^{(\alpha)}_i \overline{e^{(\alpha)}_b} = 0$. From this condition, we obtain the connection compatible with the metric:

$$h^{d\bar{b}}_{(\alpha)} (\partial_i h^{(\alpha)}_{a\bar{b}}) = h^{d\bar{b}}_{(\alpha)} (\omega^{(\alpha)}_i)^c_a h^{(\alpha)}_{c\bar{b}} = \delta^d_c (\omega^{(\alpha)}_i)^c_a = (\omega^{(\alpha)}_i)^d_a \Longleftrightarrow (\omega^{(\alpha)}_i)^a_b = h^{a\bar{c}}_{(\alpha)} (\partial_i h^{(\alpha)}_{b\bar{c}}) \tag{2.32}$$

Let us evaluate a curvature tensor associated with this connection. It is a rank two matrix-valued tensor defined as a commutator of covariant derivatives:

$$\begin{aligned}
(R^{(\alpha)}_{ij})^a_b e^{(\alpha)}_a &= -(R^{(\alpha)}_{ji})^a_b e^{(\alpha)}_a = (\nabla^{(\alpha)}_i \nabla^{(\alpha)}_j - \nabla^{(\alpha)}_j \nabla^{(\alpha)}_i) e^{(\alpha)}_b, \\
(R^{(\alpha)}_{i\bar{j}})^a_b e^{(\alpha)}_a &= -(R^{(\alpha)}_{\bar{j}i})^a_b e^{(\alpha)}_a = (\nabla^{(\alpha)}_i \nabla^{(\alpha)}_{\bar{j}} - \nabla^{(\alpha)}_{\bar{j}} \nabla^{(\alpha)}_i) e^{(\alpha)}_b, \\
(R^{(\alpha)}_{\bar{i}\bar{j}})^a_b e^{(\alpha)}_a &= -(R^{(\alpha)}_{\bar{j}\bar{i}})^a_b e^{(\alpha)}_a = (\nabla^{(\alpha)}_{\bar{i}} \nabla^{(\alpha)}_{\bar{j}} - \nabla^{(\alpha)}_{\bar{j}} \nabla^{(\alpha)}_{\bar{i}}) e^{(\alpha)}_b.
\end{aligned} \tag{2.33}$$

$(R^{(\alpha)}_{\bar{i}\bar{j}})^a_b = 0$ follows automatically from the condition $\nabla^{(\alpha)}_{\bar{i}} e^{(\alpha)}_b = 0$. Next, we take a closer look at $(R^{(\alpha)}_{ij})^a_b$. $\nabla^{(\alpha)}_i \nabla^{(\alpha)}_j e^{(\alpha)}_b$ is explicitly computed as follows:

$$\begin{aligned}
\nabla^{(\alpha)}_i \nabla^{(\alpha)}_j e^{(\alpha)}_b &= \nabla^{(\alpha)}_i ((\omega^{(\alpha)}_j)^c_b e^{(\alpha)}_c) = \nabla^{(\alpha)}_i (h^{c\bar{d}}_{(\alpha)} (\partial_j h^{(\alpha)}_{b\bar{d}}) e^{(\alpha)}_c) \\
&= (\partial_i h^{c\bar{d}}_{(\alpha)})(\partial_j h^{(\alpha)}_{b\bar{d}}) e^{(\alpha)}_c + h^{c\bar{d}}_{(\alpha)} (\partial_i \partial_j h^{(\alpha)}_{b\bar{d}}) e^{(\alpha)}_c + h^{c\bar{d}}_{(\alpha)} (\partial_j h^{(\alpha)}_{b\bar{d}})(\nabla^{(\alpha)}_i e^{(\alpha)}_c) \\
&= (\partial_i h^{a\bar{d}}_{(\alpha)})(\partial_j h^{(\alpha)}_{b\bar{d}}) e^{(\alpha)}_a + h^{a\bar{d}}_{(\alpha)} (\partial_i \partial_j h^{(\alpha)}_{b\bar{d}}) e^{(\alpha)}_a + h^{c\bar{d}}_{(\alpha)} (\partial_j h^{(\alpha)}_{b\bar{d}}) h^{a\bar{e}}_{(\alpha)} (\partial_i h^{(\alpha)}_{c\bar{e}}) e^{(\alpha)}_a.
\end{aligned}$$

Therefore, we obtain

$$\begin{aligned}
(R^{(\alpha)}_{ij})^a_b = &(\partial_i h^{a\bar{d}}_{(\alpha)})(\partial_j h^{(\alpha)}_{b\bar{d}}) + h^{a\bar{e}}_{(\alpha)} (\partial_i h^{(\alpha)}_{c\bar{e}}) h^{c\bar{d}}_{(\alpha)} (\partial_j h^{(\alpha)}_{b\bar{d}}) \\
&-(\partial_j h^{a\bar{d}}_{(\alpha)})(\partial_i h^{(\alpha)}_{b\bar{d}}) - h^{a\bar{e}}_{(\alpha)} (\partial_j h^{(\alpha)}_{c\bar{e}}) h^{c\bar{d}}_{(\alpha)} (\partial_i h^{(\alpha)}_{b\bar{d}}).
\end{aligned} \tag{2.34}$$

From this expression, we can easily show that $(R^{(\alpha)}_{ij})^a_b$ vanishes by using the equality:

$$(\partial_i h^{a\bar{b}}_{(\alpha)})h^{(\alpha)}_{c\bar{b}} + h^{a\bar{b}}_{(\alpha)}(\partial_i h^{(\alpha)}_{c\bar{b}}) = 0 \Longleftrightarrow (\partial_i h^{a\bar{b}}_{(\alpha)})h^{(\alpha)}_{c\bar{b}} = -h^{a\bar{b}}_{(\alpha)}(\partial_i h^{(\alpha)}_{c\bar{b}}), \tag{2.35}$$

obtained from differentiating $h^{a\bar{b}}_{(\alpha)}h^{(\alpha)}_{c\bar{b}} = \delta^c_a$ by $z^i_{(\alpha)}$. Lastly, we compute $(R^{(\alpha)}_{i\bar{j}})^a_b$ explicitly. Since $\nabla^{(\alpha)}_{\bar{j}} e^{(\alpha)}_b = 0$, we have

$$\begin{aligned}(R^{(\alpha)}_{i\bar{j}})^a_b e^{(\alpha)}_a &= (\nabla^{(\alpha)}_i \nabla^{(\alpha)}_{\bar{j}} - \nabla^{(\alpha)}_{\bar{j}} \nabla^{(\alpha)}_i) e^{(\alpha)}_b = -\nabla^{(\alpha)}_{\bar{j}} (\nabla^{(\alpha)}_i e^{(\alpha)}_b) \\ &= -\nabla^{(\alpha)}_{\bar{j}} (h^{c\bar{d}}_{(\alpha)} (\partial_i h^{(\alpha)}_{b\bar{d}}) e^{(\alpha)}_c) = -\partial_{\bar{j}} (h^{a\bar{d}}_{(\alpha)} (\partial_i h^{(\alpha)}_{b\bar{d}})) e^{(\alpha)}_a. \end{aligned} \tag{2.36}$$

In this way, we obtain the following expression:

$$(R^{(\alpha)}_{i\bar{j}})^a_b = -\partial_{\bar{j}} (h^{a\bar{d}}_{(\alpha)} (\partial_i h^{(\alpha)}_{b\bar{d}})). \tag{2.37}$$

Let us summarize discussions so far in the following theorem.

Theorem 2.2.1 *Let E be a holomorphic vector bundle on a complex manifold M and $(\omega^{(\alpha)}_i)^b_a$ be a connection of E compatible with Hermitian metric $h^{(\alpha)}_{a\bar{b}}$ of E. Then the curvature tensor obtained from the connection satisfies the following equalities:*

$$(R^{(\alpha)}_{ij})^a_b = (R^{(\alpha)}_{\bar{i}\bar{j}})^a_b = 0, \quad (R^{(\alpha)}_{i\bar{j}})^a_b = -(R^{(\alpha)}_{\bar{j}i})^a_b = -\partial_{\bar{j}} (h^{a\bar{d}}_{(\alpha)} (\partial_i h^{(\alpha)}_{b\bar{d}})). \tag{2.38}$$

We have obtained the local expression of the curvature tensor on U_α. Now, let us observe the relation between $(R^{(\alpha)}_{i\bar{j}})^a_b$ and $(R^{(\beta)}_{i\bar{j}})^a_b$ in $U_\alpha \cap U_\beta \neq \emptyset$. For this purpose, it is convenient to use the fact that the matrices $R^{(\alpha)}_{i\bar{j}}$ and $R^{(\beta)}_{i\bar{j}}$ are given by $-\partial_{\bar{j}}\omega^{(\alpha)}_i$ and $-\partial_{\bar{j}}\omega^{(\beta)}_i$ respectively. We differentiate (2.27) by $\bar{z}^j_{(\alpha)}$ and obtain

$$\partial_{\bar{j}}(\omega^{(\alpha)}_i) = (g^{-1}_{\beta\alpha})(\partial_{\bar{j}}\omega^{(\beta)}_i)(g_{\beta\alpha}), \tag{2.39}$$

where we used $\partial_{\bar{j}} g_{\beta\alpha} = \partial_{\bar{j}} g^{-1}_{\beta\alpha} = 0$. The above equation tells us that

$$R^{(\alpha)}_{i\bar{j}} = (g^{-1}_{\beta\alpha})(R^{(\beta)}_{i\bar{j}})(g_{\beta\alpha}). \tag{2.40}$$

In (2.40), we used local holomorphic coordinates in U_α.

As the last topic in this section, we introduce the covariant derivative of the curvature tensor $R^{(\alpha)}_{i\bar{j}}$. It is defined as follows:

$$\nabla_i (R^{(\alpha)}_{j\bar{k}})^a_b := \partial_i (R^{(\alpha)}_{j\bar{k}})^a_b - (R^{(\alpha)}_{j\bar{k}})^a_c h^{c\bar{d}}_{(\alpha)} (\partial_i h^{(\alpha)}_{b\bar{d}}) + h^{a\bar{d}}_{(\alpha)} (\partial_i h^{(\alpha)}_{c\bar{d}}) (R^{(\alpha)}_{j\bar{k}})^c_b$$
$$= \partial_i (R^{(\alpha)}_{j\bar{k}})^a_b - (R^{(\alpha)}_{j\bar{k}})^a_c (\omega^{(\alpha)}_i)^c_b + (\omega^{(\alpha)}_i)^a_c (R^{(\alpha)}_{j\bar{k}})^c_b,$$
$$\nabla_{\bar{i}} (R^{(\alpha)}_{j\bar{k}})^a_b := \partial_{\bar{i}} (R^{(\alpha)}_{j\bar{k}})^a_b. \tag{2.41}$$

We show an important identity with respect to this covariant derivative.

Theorem 2.2.2 (Bianchi's identity) *In the setting of Theorem 2.2.1, the following identities hold:*

$$\nabla_i (R^{(\alpha)}_{j\bar{k}})^a_b - \nabla_j (R^{(\alpha)}_{i\bar{k}})^a_b = 0, \qquad \nabla_{\bar{i}} (R^{(\alpha)}_{j\bar{k}})^a_b - \nabla_{\bar{k}} (R^{(\alpha)}_{j\bar{i}})^a_b = 0. \tag{2.42}$$

Proof The second identity is obvious from Theorem 2.2.1 and the definition of the covariant derivative. As for the first identity, we explicitly compute the r.h.s. in matrix form and obtain

$$\nabla_i R^{(\alpha)}_{j\bar{k}} - \nabla_j R^{(\alpha)}_{i\bar{k}} = \partial_i R^{(\alpha)}_{j\bar{k}} - R^{(\alpha)}_{j\bar{k}} \omega^{(\alpha)}_i + \omega^{(\alpha)}_i R^{(\alpha)}_{j\bar{k}} - \partial_j R^{(\alpha)}_{i\bar{k}} + R^{(\alpha)}_{i\bar{k}} \omega^{(\alpha)}_j - \omega^{(\alpha)}_j R^{(\alpha)}_{i\bar{k}}$$
$$= -\partial_i \partial_{\bar{k}} \omega^{(\alpha)}_j + (\partial_{\bar{k}} \omega^{(\alpha)}_j) \omega^{(\alpha)}_i - \omega^{(\alpha)}_i (\partial_{\bar{k}} \omega^{(\alpha)}_j)$$
$$+ \partial_j \partial_{\bar{k}} \omega^{(\alpha)}_i - (\partial_{\bar{k}} \omega^{(\alpha)}_i) \omega^{(\alpha)}_j + \omega^{(\alpha)}_j (\partial_{\bar{k}} \omega^{(\alpha)}_i)$$
$$= -\partial_{\bar{k}} (\partial_i \omega^{(\alpha)}_j - \partial_j \omega^{(\alpha)}_i + \omega^{(\alpha)}_i \omega^{(\alpha)}_j - \omega^{(\alpha)}_j \omega^{(\alpha)}_i), \tag{2.43}$$

where we used the relation $R^{(\alpha)}_{i\bar{j}} = -\partial_{\bar{j}} \omega^{(\alpha)}_i$. In the last expression of (2.43), $\partial_i \omega^{(\alpha)}_j - \partial_j \omega^{(\alpha)}_i + \omega^{(\alpha)}_i \omega^{(\alpha)}_j - \omega^{(\alpha)}_j \omega^{(\alpha)}_i$ appear, but it is nothing but $R^{(\alpha)}_{ij}$. Since $R^{(\alpha)}_{ij} = 0$, $\nabla_i R^{(\alpha)}_{j\bar{k}} - \nabla_j R^{(\alpha)}_{i\bar{k}} = 0$ follows.

2.3 Chern Classes

In the previous section, we introduced the rank r holomorphic vector bundle E on the n-dimensional complex manifold M and discussed its connection compatible with the Hermitian metric $h^{(\alpha)}_{a\bar{b}}$ on E and its curvature tensor. In this section, we introduce Chern classes defined from the curvature tensor. First, we consider matrix-valued 2-form $R^{(\alpha)} := R^{(\alpha)}_{i\bar{j}} dz^i_{(\alpha)} \wedge \overline{dz^j_{(\alpha)}}$ obtained from the curvature tensor $R^{(\alpha)}_{i\bar{j}}$ in the previous section. Since the curvature tensor is an anti-symmetric tensor, it is well-defined on U_α and satisfies

$$R^{(\alpha)} = g^{-1}_{\beta\alpha} R^{(\beta)} g_{\beta\alpha} \tag{2.44}$$

on $U_\alpha \cap U_\beta \neq \emptyset$, as can be seen from (2.40). At this stage, let us consider $\det(I_r + t\frac{i}{2\pi}R^{(\alpha)})$, where I_r is the rank r unit matrix and t is a formal variable. If we expand it in the following form:

$$\det\left(I_r + t\frac{i}{2\pi}R^{(\alpha)}\right) = 1 + tc_1(R^{(\alpha)}) + t^2c_2(R^{(\alpha)}) + \cdots + t^r c_r(R^{(\alpha)}) = \sum_{j=1} t^j c_j(R^{(\alpha)}), \tag{2.45}$$

$c_j(R^{(\alpha)})$ turns out to be a (j, j)-form on U_α. Since we have the relation:

$$\begin{aligned}\det\left(I_r + t\frac{i}{2\pi}R^{(\alpha)}\right) &= \det\left(I_r + t\frac{i}{2\pi}g_{\beta\alpha}^{-1}R^{(\beta)}g_{\beta\alpha}\right)\\ = \det\left(g_{\beta\alpha}^{-1}\left(I_r + t\frac{i}{2\pi}R^{(\beta)}\right)g_{\beta\alpha}\right) &= \det\left(I_r + t\frac{i}{2\pi}R^{(\beta)}\right),\end{aligned} \tag{2.46}$$

on $U_\alpha \cap U_\beta \neq \emptyset$, we can see that $c_j(R^{(\alpha)}) = c_j(R^{(\beta)})$ holds. Therefore, $c_j(R^{(\alpha)})$ defines a global (j, j)-form $c_j(R)$ on M. Let us reveal some interesting properties of $c_j(R)$.

For preparation, we simplify discussions of connections and curvature by introducing matrix-valued differential forms. So far, we have discussed the curvature tensor obtained from a connection compatible with the Hermitian metric, but here we discuss the curvature tensor derived from a general connection of a holomorphic vector bundle. The connections $\omega_i^{(\alpha)}$ and $\omega_{\bar{i}}^{(\alpha)}$ are organized into matrix-valued 1-forms,

$$\omega^{(\alpha)} := \omega_i^{(\alpha)}dz_{(\alpha)}^i + \omega_{\bar{i}}^{(\alpha)}d\overline{z_{(\alpha)}^i} = \omega_i^{(\alpha)}dz_{(\alpha)}^i, \tag{2.47}$$

where we used $\omega_{\bar{i}}^{(\alpha)} = 0$, which holds in the case of a holomorphic vector bundle. We then introduce the curvature form $R^{(\alpha)} = \frac{1}{2}R_{ij}^{(\alpha)}dz_{(\alpha)}^i \wedge dz_{(\alpha)}^j + R_{i\bar{j}}^{(\alpha)}dz_{(\alpha)}^i \wedge d\overline{z_{(\alpha)}^j}$. Note that we include $\frac{1}{2}R_{ij}^{(\alpha)}dz_{(\alpha)}^i \wedge dz_{(\alpha)}^j$ since the condition $R_{ij}^{(\alpha)} = 0$ does not hold in the case of a general connection. ($R_{\bar{i}\bar{j}}^{(\alpha)}$ vanishes trivially from the condition $\omega_{\bar{i}}^{(\alpha)} = 0$.) With this setting, the relation between the curvature form and the connection form is simplified into the following:

$$R^{(\alpha)} = d\omega^{(\alpha)} + \omega^{(\alpha)} \wedge \omega^{(\alpha)}. \tag{2.48}$$

If we evaluate the r.h.s. by using the exterior differential operator $d = dz_{(\alpha)}^i \wedge \partial_i + d\overline{z_{(\alpha)}^i} \wedge \partial_{\bar{i}}$, we can confirm the results in the previous section:

$$\begin{aligned}
&d\omega^{(\alpha)} + \omega^{(\alpha)} \wedge \omega^{(\alpha)} \\
&= (\partial_i \omega_j^{(\alpha)}) dz_{(\alpha)}^i \wedge dz_{(\alpha)}^j + (\partial_{\bar{j}} \omega_i^{(\alpha)}) \overline{dz_{(\alpha)}^j} \wedge dz_{(\alpha)}^i + (\omega_i^{(\alpha)} dz_{(\alpha)}^i) \wedge (\omega_j^{(\alpha)} dz_{(\alpha)}^j) \\
&= \frac{1}{2}(\partial_i \omega_j^{(\alpha)} - \partial_j \omega_i^{(\alpha)} + \omega_i^{(\alpha)} \omega_j^{(\alpha)} - \omega_j^{(\alpha)} \omega_i^{(\alpha)}) dz_{(\alpha)}^i \wedge dz_{(\alpha)}^j + (-\partial_{\bar{j}} \omega_i^{(\alpha)}) dz_{(\alpha)}^i \wedge \overline{dz_{(\alpha)}^j} \\
&= \frac{1}{2} R_{ij}^{(\alpha)} dz_{(\alpha)}^i \wedge dz_{(\alpha)}^j + R_{i\bar{j}}^{(\alpha)} dz_{(\alpha)}^i \wedge \overline{dz_{(\alpha)}^j}. \qquad (2.49)
\end{aligned}$$

If we use (2.48), we can derive Bianchi's identity easily in the following way:

$$\begin{aligned}
dR^{(\alpha)} &= d^2 \omega^{(\alpha)} + (d\omega^{(\alpha)}) \wedge \omega^{(\alpha)} - \omega^{(\alpha)} \wedge (d\omega^{(\alpha)}) \\
&= (d\omega^{(\alpha)} + \omega^{(\alpha)} \wedge \omega^{(\alpha)}) \wedge \omega^{(\alpha)} - \omega^{(\alpha)} \wedge (d\omega^{(\alpha)} + \omega^{(\alpha)} \wedge \omega^{(\alpha)}) \\
&= R^{(\alpha)} \wedge \omega^{(\alpha)} - \omega^{(\alpha)} \wedge R^{(\alpha)}. \qquad (2.50)
\end{aligned}$$

This shows how BIanchi's identity is extended to the case when $R_{ij}^{(\alpha)}$ does not vanish:

$$\begin{aligned}
\nabla_i R_{j\bar{k}}^{(\alpha)} - \nabla_j R_{i\bar{k}}^{(\alpha)} + \nabla_{\bar{k}} R_{ij}^{(\alpha)} &= 0 \\
\nabla_{\bar{i}} R_{j\bar{k}}^{(\alpha)} - \nabla_{\bar{k}} R_{j\bar{i}}^{(\alpha)} &= 0. \qquad (2.51)
\end{aligned}$$

With this set-up, we derive some properties of $c_j(R)$. First of all **$c_j(R)$ is a closed form, i.e., $dc_j(R) = 0$.** To show this, we have to note the following equality:

$$\begin{aligned}
\det\left(I_r + t\frac{i}{2\pi}R\right) &= \exp\left(\mathrm{tr}\left(\log\left(I_r + t\frac{i}{2\pi}R\right)\right)\right) = \exp\left(\mathrm{tr}\left(\sum_{j=1}^{\infty} \frac{(-1)^{j-1}}{j} t^j \left(\frac{i}{2\pi}\right)^j R^j\right)\right) \\
&= \exp\left(\sum_{j=1}^{\infty} \frac{(-1)^{j-1}}{j} t^j \left(\frac{i}{2\pi}\right)^j \mathrm{tr}(R^j)\right). \qquad (2.52)
\end{aligned}$$

Here, we omit the subscripts (α) and $\wedge$ for brevity, and we use this notation for some time to come. From (2.52), we can see that $c_j(R)$ is represented as a polynomial in $u_k(R) := \mathrm{tr}(R^k)$. Therefore, it is enough for us to show $du_k(R) = 0$ for $k \geq 1$. This can be derived as follows. Since we have $dR = R\omega - \omega R$ in (2.50),

$$\begin{aligned}
du_k(R) &= \mathrm{tr}\left(\sum_{j=0}^{k-1} R^j (R\omega - \omega R) R^{k-1-j}\right) = \mathrm{tr}\left(\sum_{j=0}^{k-1} R^{j+1} \omega R^{k-1-j} - R^j \omega R^{k-j}\right) \\
&= \sum_{j=0}^{k-1} (\mathrm{tr}(R^{j+1} \omega R^{k-1-j}) - \mathrm{tr}(R^j \omega R^{k-j})) = \sum_{j=0}^{k-1} (\mathrm{tr}(R^k \omega) - \mathrm{tr}(R^k \omega)) = 0. \qquad (2.53)
\end{aligned}$$

Here we used the fact that the trace is invariant under cyclic permutation of the order of the product.

Since $c_j(R)$ turns out to be a closed form, it determines a cohomology class $[c_j(R)] \in H^*(M)$. From now on, we show that the cohomology class $[c_j(R)]$ does not depend on connection ω. In other words, it is determined only from the vector bundle E. To show this, we use again the closed form $u_k(R) = \mathrm{tr}(R^k)$. Let ω and $\tilde{\omega}$ be two different connections of E and R and $\tilde{R}$ be curvatures obtained from these two connections. Then it is enough for us to show that there exists a global differential form θ which satisfies

$$u_k(R) - u_k(\tilde{R}) = d\theta. \tag{2.54}$$

First, we introduce a family of connections $\omega_t = t\omega + (1-t)\tilde{\omega}$ which connects ω and $\tilde{\omega}$. Note here that $\gamma := \partial_t \omega_t = \omega - \tilde{\omega}$ satisfies

$$\gamma^{(\alpha)} = g_{\beta\alpha}^{-1} \gamma_t^{(\beta)} g_{\beta\alpha}. \tag{2.55}$$

Let R_t be a family of curvatures obtained from the connection ω_t. By differentiating both sides of the following relation:

$$R_t = d\omega_t + \omega_t \omega_t, \tag{2.56}$$

by t, we obtain

$$\partial_t R_t = d\gamma + \gamma\omega_t + \omega_t\gamma. \tag{2.57}$$

Using this formula, we evaluate $\partial_t u_k(R_t)$:

$$\begin{aligned} \partial_t u_k(R_t) &= \partial_t(\mathrm{tr}(R_t^k)) = \sum_{j=0}^{k-1} \mathrm{tr}(R_t^j (\partial_t R_t) R_t^{k-1-j}) \\ &= k\mathrm{tr}((\partial_t R_t) R_t^{k-1}) = k\mathrm{tr}((d\gamma + \gamma\omega_t + \omega_t\gamma) R_t^{k-1}). \end{aligned} \tag{2.58}$$

On the other hand, $\mathrm{tr}(\gamma R_t^{k-1})$ turns out to be a global form on M since (2.55) holds. We evaluate $d\mathrm{tr}(\gamma R_t^{k-1})$ as follows:

$$\begin{aligned} d\mathrm{tr}(\gamma R_t^{k-1}) &= \mathrm{tr}(d\gamma R_t^{k-1}) - \sum_{j=0}^{k-2} \mathrm{tr}(\gamma R_t^j (dR_t) R_t^{k-2-j}) \\ &= \mathrm{tr}(d\gamma R_t^{k-1}) - \sum_{j=0}^{k-2} \mathrm{tr}(\gamma R_t^j (R_t\omega_t - \omega_t R_t) R_t^{k-2-j}) \\ &= \mathrm{tr}(d\gamma R_t^{k-1}) - \sum_{j=0}^{k-2} \mathrm{tr}(\gamma R_t^{j+1} \omega_t R_t^{k-2-j}) + \sum_{j=0}^{k-2} \mathrm{tr}(\gamma R_t^j \omega_t R_t^{k-1-j}) \\ &= \mathrm{tr}(d\gamma R_t^{k-1}) - \mathrm{tr}(\gamma R_t^{k-1}\omega_t) + \mathrm{tr}(\gamma\omega_t R_t^{k-1}) \\ &= \mathrm{tr}(d\gamma R_t^{k-1}) + \mathrm{tr}(\omega_t\gamma R_t^{k-1}) + \mathrm{tr}(\gamma\omega_t R_t^{k-1}) = \mathrm{tr}((d\gamma + \gamma\omega_t + \omega_t\gamma) R_t^{k-1}). \end{aligned} \tag{2.59}$$

In this derivation, we used Bianchi's identity $dR_t = R_t\omega_t - \omega_t R_t$. We have to note that $\mathrm{tr}(\gamma R_t^{k-1}\omega_t) = -\mathrm{tr}(\omega_t\gamma R_t^{k-1})$ since both ω_t and γ are matrix-valued 1-form. Combination of (2.58) and (2.59) gives us

$$\partial_t u_k(R_t) = kd\mathrm{tr}(\gamma R_t^{k-1}) = d\mathrm{tr}(k\gamma R_t^{k-1}). \tag{2.60}$$

Integration of both sides of the above equation from $t = 0$ to $t = 1$ yields the following equation.

$$\int_0^1 dt\partial_t u_k(R_t) = u_k(R_1) - u_k(R_0) = u_k(\tilde{R}) - u_k(R) = \int_0^1 dt\left(d\mathrm{tr}(k\gamma R_t^{k-1})\right) = d\mathrm{tr}(k\int_0^1 dt(\gamma R_t^{k-1})). \tag{2.61}$$

In this way, we obtain $u_k(\tilde{R}) - u_k(R) = d\theta$ by setting $\mathrm{tr}(k\int_0^1 dt(\gamma R_t^{k-1})$ as θ. Let us summarize the discussions so far as the following theorem.

Theorem 2.3.1 *Let E be a rank r holomorphic vector bundle on a complex manifold and R be a curvature obtained from a connection ω of E. The global $2j$-form $c_j(R)$ on M defined by*

$$det\left(I_r + t\frac{i}{2\pi}R^{(\alpha)}\right) = \sum_{j=0}^{r} t^r c_j(R), \tag{2.62}$$

is a closed form and its cohomology class $[c_j(R)]$ does not depend on ω and is determined only by E. Therefore we define $c_j(E)$ by $[c_j(R)]$ and call it the j-th Chern class of E. We call the cohomology class give by

$$c(E) := [\sum_{j=0}^{r} t^r c_j(R)] = \sum_{j=0}^{r} t^r c_j(E), \tag{2.63}$$

the total Chern class of E and $c_r(E)$ the top Chern class of E. If we take ω as the connection compatible with the Hermitian metric of E, $c_j(R)$ becomes a (j, j) form.

2.3.1 Calculus of Holomorphic Vector Bundles and Chern Classes

In this subsection, we introduce calculus defined among holomorphic vector bundles on a complex manifold and discuss how to compute Chern classes of the holomorphic vector bundles obtained from the calculus. We consider holomorphic vector bundles with a Hermitian metric and assume that connections and curvatures are compatible with the Hermitian metric.

Let us define the pull-back of a vector bundle. We consider a holomorphic map f from an n-dimensional complex manifold N to an m-dimensional complex manifold M. f is a holomorphic map if for any holomorphic coordinate system $\phi_N : U_N \to \mathbf{C}^n$ and $\psi_M : U_M \to \mathbf{C}^m$ with $f(U_N) \cap U_M \neq$, $\psi_M \circ f \circ \phi_N^{-1}$ is holomorphic. Let $\pi_E : E \to M$ be a rank r holomorphic vector bundle. Roughly speaking, a rank r holomorphic vector bundle $f^{-1}E$ on N is defined by attaching the fiber vector space $\pi_E^{-1}(f(x))$ as the fiber of $x \in N$. We take a closer look at this definition in terms of the transition function of a vector bundle. Let $\bigcup_{\alpha} U_\alpha$ be an open covering used in local trivialization of E and $g_{\beta\alpha} : U_\alpha \cap U_\beta \to GL(r, \mathbf{C})$ be the transition function obtained from the trivialization. Then $f^{-1}E$ is defined as a vector bundle whose open coverings used in local trivialization and transition functions are $\bigcup_{\alpha} f^{-1}(U_\alpha)$ and $f^* g_{\beta\alpha} : f^{-1}(U_\alpha) \cap f^{-1}(U_\beta) \to GL(r, \mathbf{C})$ respectively. Here $f^* g_{\beta\alpha}$ is a pull-back function defined by $f^* g_{\beta\alpha}(x) = g_{\beta\alpha}(f(x))$.

Then, how are Chern classes of N of $f^{-1}E$ related to Chern classes of E? We have the following simple result for this question.

Theorem 2.3.2

$$c_j(f^{-1}E) = f^* c_j(E), \tag{2.64}$$

where f^* is the homomorphism of cohomology rings $f^* : H^*(M) \to H^*(N)$ induced by pull-back of differential forms.

We omit the detailed proof, but from the above discussion of the transition function of $f^{-1}E$, we can easily see that the Hermitian metric of $f^{-1}E$ naturally induced from the Hermitian metric $h_{a\bar{b}}$ of E is given by $f^* h_{a\bar{b}}$. Therefore, for the curvature tensor R obtained from $h_{a\bar{b}}$, the curvature tensor of $f^{-1}E$ is evaluated as $f^* R$. Then the assertion of the theorem is obvious.

Next, we consider a rank r holomorphic vector bundle $\pi_E : E \to M$ and a rank s holomorphic vector bundle $\pi_F : F \to M$ on an m-dimensional complex manifold M. By attaching the fiber vector space $\pi_E^{-1}(x) \oplus \pi_F^{-1}(x)$ to $x \in M$, we can construct a rank $r + s$ holomorphic vector bundle $E \oplus F$ on M. For simplicity. we take a common open covering $\bigcup_{\alpha} U_\alpha$ for local trivialization of both E and F. If we denote the transition functions of E and F by $g^E_{\beta\alpha}$ and F $g^F_{\beta\alpha}$ respectively, transition function of $E \oplus F$ is given by $g^E_{\beta\alpha} \oplus g^F_{\beta\alpha} : U_\alpha \cap U_\beta \to GL(r + s, \mathbf{C})$. Here, $g^E_{\beta\alpha} \oplus g^F_{\beta\alpha}$ is explicitly written in matrix form as follows:

$$g^E_{\beta\alpha} \oplus g^F_{\beta\alpha} := \begin{pmatrix} g^E_{\beta\alpha} & \mathbf{0} \\ \mathbf{0} & g^F_{\beta\alpha} \end{pmatrix}. \tag{2.65}$$

Then, how are Chern classes of $E \oplus F$ related to Chern classes of E and F? We have beautiful answer for this question.

Theorem 2.3.3

$$c(E \oplus F) = \sum_{j=0}^{r+s} t^j c_j(E \oplus F) = c(E)c(F) = \Big(\sum_{j=0}^{r} t^j c_j(E)\Big)\Big(\sum_{j=0}^{s} t^j c_j(F)\Big). \tag{2.66}$$

By expanding the r.h.s. in t, we can represent $c_n(E \oplus F)$ as a polynomial of $c_i(E)$ and $c_j(E)$. Let us present a brief sketch of the proof. If we denote the curvatures of E and F by R_E and R_F respectively, the curvature of $E \oplus F$ is given by $R_E \oplus R_F$. Therefore, the following equality holds:

$$\begin{aligned}\sum_{j=0}^{r+s} t^j c_j(R_E \oplus R_F) &= \det\left(I_{r+s} + t\frac{i}{2\pi}(R_E \oplus R_F)\right) = \begin{vmatrix} I_r + t\frac{i}{2\pi}R_E & \mathbf{0} \\ \mathbf{0} & I_s + t\frac{i}{2\pi}R_F \end{vmatrix} \\ &= \det\left(I_r + t\frac{i}{2\pi}R_E\right)\det\left(I_s + t\frac{i}{2\pi}R_F\right) \\ &= \Big(\sum_{j=0}^{r} t^j c_j(R_E)\Big)\Big(\sum_{j=0}^{s} t^j c_j(R_F)\Big). \end{aligned} \tag{2.67}$$

The assertion of the theorem follows from this equality.

On the other hand, we can construct a rank rs holomorphic vector bundle $E \otimes F$ by attaching the vector space $\pi_E^{-1}(x) \otimes \pi_F^{-1}(x)$ to $x \in M$. Its transition function is given by $g^E_{\beta\alpha} \otimes g^F_{\beta\alpha} : U_\alpha \cap U_\beta \to GL(rs, \mathbf{C})$ where $g^E_{\beta\alpha} \otimes g^F_{\beta\alpha}$ is the tensor product of matrices. Then how can we write Chern classes of $E \otimes F$ in terms of Chern classes of E and F? Let us consider first the case when $r = s = 1$. A rank 1 holomrphic vector bundle is called a complex line bundle and it plays important roles in the theory of complex manifolds. When E and F are line bundles, $E \otimes F$ is also a line bundle. Therefore its non-trivial Chern class is given by the first Chern class. We have the following formula for its first Chern class.

Theorem 2.3.4

$$c_1(E \otimes F) = c_1(E) + c_1(F). \tag{2.68}$$

We give here a sketch of the proof. Let R_E and R_F be curvatures of E and F respectively. Since $r = s = 1$, curvature tensors are given by complex scalar-valued 2-forms. Therefore, $c_1(R_E)$ and $c_1(R_F)$ are given by $\frac{i}{2\pi}R_E$ and $\frac{i}{2\pi}R_F$ respectively. Let h_E and h_F be Hermitian metrics of E and F. These are also locally complex scalar-valued functions. By using the results obtained so far, we can write down connections compatible with these Hermitian metrics as $\omega_E = (h_E)^{-1}\partial_i h_E dz^i = (\partial_i \log(h_E))dz^i$ and $\omega_F = (h_F)^{-1}\partial_i h_F dz^i = (\partial_i \log(h_F))dz^i$. Curvatures obtained from these metrics are also written as $R_E = -(\partial_{\bar{j}}\partial_i \log(h_E))dz^i \wedge \overline{dz^j}$ and $R_F = -(\partial_{\bar{j}}\partial_i \log(h_F))dz^i \wedge \overline{dz^j}$. Therefore, we can write down the first Chern classes of E and F in terms of their Hermitian metrics:

$$c_1(R_E) = -\frac{i}{2\pi}(\partial_{\bar{j}}\partial_i \log(h_E))dz^i \wedge \overline{dz^j},$$
$$c_1(R_F) = -\frac{i}{2\pi}(\partial_{\bar{j}}\partial_i \log(h_F))dz^i \wedge \overline{dz^j}. \tag{2.69}$$

On the other hand, the Hermitian metric $h_{E\otimes F}$ of $E \otimes F$ is given by $h_E h_F$ and $c_1(R_{E\otimes F})$ is given by $-(\partial_{\bar{j}}\partial_i \log(h_{E\otimes F}))dz^i \wedge \overline{dz^j}$. Then the assertion of the theorem is obvious.

As the last topic of this subsection, we discuss Chern classes of a dual vector bundle E^* of a rank r holomorphic vector bundle E on a complex manifold M. E^* is the vector bundle whose fiber of $x \in M$ is given by the dual vector space $(\pi_E^{-1}(x))^*$, and its transition function is written as ${}^t(g_{\beta\alpha}^{-1})$ where $g_{\beta\alpha}$ is the one for E. Let us consider the case when E is a line bundle, i.e., $r = 1$. In this case, the transition function of E takes a complex scalar value. Hence the transition function of E^* is given by $g_{\beta\alpha}^{-1} = \frac{1}{g_{\beta\alpha}}$. Therefore, if we denote the Hermitian metric of E by h_E, we can set the Hermitian metric of E^* as $h_E^{-1} = \frac{1}{h_E}$. Then we use discussions that appeared in the proof of Theorem 2.3.4 and obtain

$$c_1(R_{E^*}) = -\frac{i}{2\pi}(\partial_{\bar{j}}\partial_i \log(h_{E^*}))dz^i \wedge \overline{dz^j} = -\frac{i}{2\pi}(\partial_{\bar{j}}\partial_i \log(h_E^{-1}))dz^i \wedge \overline{dz^j}$$
$$= \frac{i}{2\pi}(\partial_{\bar{j}}\partial_i \log(h_E))dz^i \wedge \overline{dz^j} = -c_1(R_E). \tag{2.70}$$

In this way, we conclude that $c_1(E^*) = -c_1(E)$ when E is a line bundle. Then we turn to the case of a vector bundle E with rank r. In order to use the splitting principle, we virtually decompose E into the direct sum of line bundles.

$$E = e_1 \oplus e_2 \oplus \cdots \oplus e_r. \tag{2.71}$$

Accordingly, E^* is also decomposed into

$$E^* = e_1^* \oplus e_2^* \oplus \cdots \oplus e_r^*. \tag{2.72}$$

From the result of the case of a line bundle, we obtain

$$c(E^*) = (1 + tc_1(e_1^*))(1 + tc_1(e_2^*)) \cdots (1 + tc_1(e_r^*))$$
$$= (1 - tc_1(e_1))(1 - tc_1(e_2)) \cdots (1 - tc_1(e_r)). \tag{2.73}$$

Therefore, we can write down Chern classes of E^* in terms of those of E as follows.

Theorem 2.3.5

$$c_j(E^*) = (-1)^j c_j(E). \tag{2.74}$$

2.4 Kähler Manifolds and Projective Spaces

2.4.1 *Definition of Kähler Manifolds*

Let M be a complex n-dimensional manifold and $\bigcup_\alpha U_\alpha$ be an open covering of M which gives a complex coordinate system. $(z^1_{(\alpha)}, z^2_{(\alpha)}, \dots, z^n_{(\alpha)})$ is a local complex coordinate system on U_α. Let us consider a symbolic partial differential $\frac{\partial}{\partial z^j_{(\alpha)}}$. Since M is a complex manifold, $\frac{\partial \overline{z^k_{(\beta)}}}{\partial z^j_{(\alpha)}}$ vanishes. Hence the chain rule of partial differentials is given by

$$\frac{\partial}{\partial z^j_{(\alpha)}} = \frac{\partial z^k_{(\beta)}}{\partial z^j_{(\alpha)}} \frac{\partial}{\partial z^k_{(\beta)}}. \tag{2.75}$$

Since $z^k_{(\beta)}(z^j_{(\alpha)}, \dots, z^n_{(\alpha)})$ is a holomorphic function in $z^j_{(\alpha)}$ $(j = 1, \dots, n)$, $\frac{\partial z^k_{(\beta)}}{\partial z^j_{(\alpha)}}$ is also holomorphic. Therefore, by taking $\frac{\partial}{\partial z^j_{(\alpha)}}$, $(j = 1, \dots, n)$ as the local basis used for trivialization on U_α and $\frac{\partial z^k_{(\beta)}}{\partial z^j_{(\alpha)}}$ as the transition function $(g_{\beta\alpha})^k_j$, we can define a rank n holomorphic vector bundle on M. We call this vector bundle the holomorphic tangent bundle on M and denote it by $T'M$. Note here that we can take common subscripts to distinguish the local trivialization basis and local coordinates. Let $g^{(\alpha)}_{i\bar{j}}$ be an Hermitian metric of $T'M$. We define a $(1, 1)$ differential form $g := g^{(\alpha)}_{i\bar{j}} dz^i_{(\alpha)} \wedge \overline{dz^j_{(\alpha)}} = g^{(\beta)}_{i\bar{j}} dz^i_{(\beta)} \wedge \overline{dz^j_{(\beta)}}$ associated with the Hermitian metric.

Definition 2.4.1 Let M be a complex n-dimensional manifold. If the $(1, 1)$-form $g := g_{i\bar{j}} dz^i \wedge \overline{dz^j}$ associated with the Hermitian metric $g_{i\bar{j}}$ of $T'M$ satisfies the condition $dg = 0$, the tuple (M, g) is called a Kähler manifold. If (M, g) is a Kähler manifold, $g_{i\bar{j}}$ and g are called the Kähler metric and Kähler form respectively.

In the above definition, we omit the subscript (α) for U_α to avoid complexity of notation. From now on, we sometimes omit g from the tuple (M, g) by only stating that "M is a Kähler manifold." Let us take a closer look at the condition $dg = 0$:

$$\begin{aligned} dg &= \partial_k g_{i\bar{j}} dz^k \wedge dz^i \wedge d\overline{z^j} + \partial_{\bar{k}} g_{i\bar{j}} d\overline{z^k} \wedge dz^i \wedge d\overline{z^j} \\ &= \frac{1}{2}(\partial_k g_{i\bar{j}} - \partial_i g_{k\bar{j}}) dz^k \wedge dz^i \wedge d\overline{z^j} + \frac{1}{2}(\partial_{\bar{k}} g_{i\bar{j}} - \partial_{\bar{j}} g_{i\bar{k}}) dz^i \wedge d\overline{z^j} \wedge d\overline{z^k} \\ &= 0, \Longleftrightarrow \partial_k g_{i\bar{j}} = \partial_i g_{k\bar{j}}, \quad \partial_{\bar{k}} g_{i\bar{j}} = \partial_{\bar{j}} g_{i\bar{k}}. \end{aligned} \tag{2.76}$$

As in the case of general holomorphic vector bundles, we introduce a contravariant metric $g^{i\bar{j}}$ defined by

$$g^{i\bar{k}} g_{j\bar{k}} = \delta^i_j, \quad g^{k\bar{i}} g_{k\bar{j}} = \delta^{\bar{i}}_{\bar{j}}. \tag{2.77}$$

With this definition, we can construct a connection compatible with the metric by using the results in Sect. 2.2. When M is a Kähler manifold, we denote it by Γ^i_{jk}. From the previous results, we obtain

$$\Gamma^i_{jk} = g^{i\bar{l}} \partial_j g_{k\bar{l}}, \tag{2.78}$$

but from the condition (2.76), we also have the relation:

$$\Gamma^i_{jk} = \Gamma^i_{kj}. \tag{2.79}$$

Therefore, we can regard this connection as a kind of extension of the Levi–Civita connection of Riemann manifolds to complex manifolds. As for the curvature tensor, $R^i{}_{jk\bar{l}}$, which is the only non-vanishing element, is given by

$$R^i{}_{jk\bar{l}} = -\partial_{\bar{l}} \Gamma^i_{jk}, \tag{2.80}$$

but it also satisfies

$$R^i{}_{jk\bar{l}} = R^i{}_{kj\bar{l}}, \tag{2.81}$$

since (2.79) holds. This is the extension of the Riemann curvature tensor to Kähler manifolds. The Ricci curvature tensor is defined in the same way as Riemann manifolds,

$$R_{i\bar{j}} = R^k{}_{ik\bar{j}}. \tag{2.82}$$

In the case of Kähler manifolds, we obtain the following relation from (2.81):

$$R_{i\bar{j}} = R^k{}_{ik\bar{j}} = R^k{}_{ki\bar{j}}. \tag{2.83}$$

This says that **the Ricci curvature tensor of a Kähler manifold equals the trace of the Riemann curvature tensor**.

Now, let us discuss Chern classes of $T'M$. We define a matrix-valued (1, 1)-form R by $(R)^i_j = R^i_{jk\bar{l}} dz^k \wedge d\overline{z^l}$, where $R^i_{jk\bar{l}}$ is the curvature tensor defined above. Then the (j, j)-form $c_j(R)$ defined by

$$\det\left(I_n + t\frac{i}{2\pi}R\right) = \sum_{j=0}^{n} t^j c_j(R), \tag{2.84}$$

is a global closed form on M, and its cohomology class is nothing but the Chern class $c_j(T'M)$. This Chern class is called the j-th Chern class of M and is denoted by $c_j(M)$. As can be seen from discussions of the characteristic polynomial of a matrix, $c_1(R)$ in (2.84) is given by

$$c_1(R) = \frac{i}{2\pi}\mathrm{tr}(R) = \frac{i}{2\pi}R^j_{jk\bar{l}} dz^k \wedge d\overline{z^l} = \frac{i}{2\pi}R_{k\bar{l}} dz^k \wedge d\overline{z^l}. \tag{2.85}$$

Therefore, $c_1(R)$ equals the Ricci form $R_{k\bar{l}} dz^k \wedge d\overline{z^l}$ multiplied by $\frac{i}{2\pi}$. Then, the following question occurs in our mind. "If a Kähler manifold M satisfies the condition $c_1(M) = [c_1(R)] = 0$, can it have a Kähler metric $g_{i\bar{j}}$ whose curvature tensor $R_{i\bar{j}}$ vanishes at any $x \in M$?" E. Calabi conjectured that this question is affirmative for compact Kähler manifolds and the conjecture was proved by S.T. Yau.

Theorem 2.4.1 *If a compact Kähler manifold M satisfies the condition $c_1(M) = 0$, there exists a unique Kähler metric $g_{i\bar{j}}$ of M whose Ricci curvature $R_{i\bar{j}}$ vanishes at any $x \in M$.*

This is the reason why we call a compact Kähler manifold with vanishing first Chern class a Calabi–Yau manifold. We introduce here one of the modern definitions of a complex n-dimensional Calabi–Yau manifold.

Definition 2.4.2 A complex n-dimensional compact Kähler manifold M is a Calabi–Yau manifold if it satisfies the following conditions.
(i) $h^{0,1}(M) = h^{0,2} = \cdots = h^{0,n-1}(M) = 0$. (ii) $c_1(M) = 0$.

In the above definition, $h^{p,q}(M)$ is the Hodge number of M with degree (p, q) that will be introduced in the next subsection. (In the $n = 1$ case, we conventionally include a real 2-dimensional torus T^2 in Calabi–Yau manifolds.) Of course, Chern classes of M other than $c_1(M)$ are also important indices to see how $T'M$ is globally twisted or how M is topologically complicated. Especially, the following result on top Chern class $c_n(M)$ is well-known. (It is a variant of Gauss–Bonnet theorem.)

Theorem 2.4.2 *For an n-dimensional compact Kähler manifold M, the following equality holds:*

$$\int_M c_n(M) = \chi(M) = \sum_{j=0}^{2n}(-1)^j \dim_{\mathbf{R}}(H^j(M, \mathbf{R})), \tag{2.86}$$

where $\chi(M)$ is the topological Euler number of M.

2.4.2 Dolbeault Cohomology of Kähler Manifolds

Let M be a complex n-dimensional compact Kähler manifold and $\Omega^{p,q}(M)$ be an infinite-dimensional complex vector space of differential forms of bi-degree (p, q),

$$f_{i_1 i_2 \cdots i_p \bar{j}_1 \bar{j}_2 \cdots \bar{j}_q} dz^{i_1} \wedge \cdots \wedge dz^{i_p} \wedge d\overline{z^{j_1}} \wedge \cdots \wedge d\overline{z^{j_q}}, \tag{2.87}$$

on M. The exterior derivative operator d on M is decomposed into $d = dz^i \wedge \partial_i + d\overline{z^i} \wedge \partial_{\bar{i}}$. Then we define the holomorphic exterior derivative operator ∂ and the anti-holomorphic one $\overline{\partial}$ as follows:

$$\partial := dz^i \wedge \partial_i, \quad \overline{\partial} := d\overline{z^i} \wedge \partial_{\bar{i}}. \tag{2.88}$$

We can easily observe the following relations:

$$\partial^2 = 0, \quad \partial\overline{\partial} + \overline{\partial}\partial = 0, \quad \overline{\partial}^2 = 0. \tag{2.89}$$

Let $\overline{\partial}^{p,q} : \Omega^{p,q}(M) \to \Omega^{p,q+1}(M)$ be the restriction of $\overline{\partial}$ to $\Omega^{p,q}(M)$. From (2.89), $\overline{\partial}^{p,q}$: obviously satisfies $\overline{\partial}^{p,q+1} \circ \overline{\partial}^{p,q} = 0$ and yields a differential complex:

$$0 \xrightarrow{\overline{\partial}^{p,-1}} \Omega^{p,0}(M) \xrightarrow{\overline{\partial}^{p,0}} \Omega^{p,1}(M) \xrightarrow{\overline{\partial}^{p,1}} \cdots \xrightarrow{\overline{\partial}^{p,n-1}} \Omega^{p,n}(M) \xrightarrow{\overline{\partial}^{p,n}} 0. \tag{2.90}$$

It is natural to consider the cohomology $\mathrm{Ker}(\overline{\partial}^{p,q})/\mathrm{Im}(\overline{\partial}^{p,q-1})$ of the above complex.

Definition 2.4.3 We denote the cohomology $\mathrm{Ker}(\overline{\partial}^{p,q})/\mathrm{Im}(\overline{\partial}^{p,q-1})$ of the differential complex (2.90) by $H^{p,q}_{\overline{\partial}}(M)$ and call it the Dolbeault cohomology of bi-degree (p, q) of M.

Let us discuss the relation between the Dolbeault cohomology of M and the De Rham cohomology of M. By setting $\Omega^i(M) := \bigoplus_{p+q=i} \Omega^{p,q}(M)$ and $d^i : \Omega^i(M) \to \Omega^{i+1}(M)$ as the restriction of $d = \partial + \overline{\partial}$ to $\Omega^i(M)$, we can construct a differential complex,

$$0 \xrightarrow{d^{-1}} \Omega^0(M) \xrightarrow{d^0} \Omega^1(M) \xrightarrow{d^1} \cdots \xrightarrow{d^{2n-1}} \Omega^{2n}(M) \xrightarrow{d^{2n}} 0. \tag{2.91}$$

The De Rham cohomology $H^i(M, \mathbf{C})$ is nothing but the cohomology $\mathrm{Ker}(d^i)/\mathrm{Im}(d^{i-1})$ of this comolex. $\dim_{\mathbf{C}}(H^i(M, \mathbf{C}))$ is called the i-th Betti number of M and is denoted by $b^i(M)$. On the other hand, $\dim_{\mathbf{C}}(H^{p,q}_{\overline{\partial}}(M))$ is called the (p, q) Hodge number of M and denoted by $h^{p,q}(M)$. (Originally, "Hodge number" was used for the dimension of the vector space of (p, q) harmonic forms on M, but we use it in the above meaning for convenience.)

Theorem 2.4.3 *For a complex n-dimensional compact Kähler manifold M, the following relations hold:*

$$(i)\, b^i(M) = \sum_{p+q=i} h^{p,q}(M).\ (ii)\, h^{p,q}(M) = h^{q,p}(M).\ (iii)\, h^{n-p,n-q}(M) = h^{p,q}(M).$$

We omit the proof of this theorem. (i) says that the Dolbeault cohomology is a refinement of the De Rham cohomology with holomorphic degree p and anti-holomorphic degree q, (ii) says that it has symmetry under the interchange of p and q, and (iii) asserts that there exists an extension of Poincaré duality that holds for a real compact manifold to a compact Kähler manifold. (It is called Serre duality.)

As the last topic of this subsection, we introduce the Dolbeault cohomology of a holomorphic vector bundle of E on a compact Kähler manifold M. Let $\Omega^{p,q}(M, E)$ be a complex infinite-dimensional vector space of (p, q) differential forms which take values in E,

$$f^E_{i_1 i_2 \cdots i_p \bar{j}_1 \bar{j}_2 \cdots \bar{j}_q} dz^{i_1} \wedge \cdots \wedge dz^{i_p} \wedge \overline{dz^{j_1}} \wedge \cdots \wedge \overline{dz^{j_q}}. \tag{2.92}$$

(To be more precise, $f^E_{i_1 i_2 \cdots i_p \bar{j}_1 \bar{j}_2 \cdots \bar{j}_q}$ is a section of the vector bundle $E \otimes (\bigwedge^p T'M) \otimes (\bigwedge^q \overline{T'M})$.) Since E is holomorphic,the anti-holomorphic part of its connection vanishes. Therefore, $\overline{\partial} := d\overline{z^i} \wedge \partial_{\bar{i}}$ acts on $\Omega^{p,q}(M, E)$ and $\overline{\partial}^{p,q} := \overline{\partial}|_{\Omega^{p,q}(M,E)}$ yields the following differential complex:

$$0 \xrightarrow{\overline{\partial}^{p,-1}} \Omega^{p,0}(M, E) \xrightarrow{\overline{\partial}^{p,0}} \Omega^{p,1}(M, E) \xrightarrow{\overline{\partial}^{p,1}} \cdots \xrightarrow{\overline{\partial}^{p,n-1}} \Omega^{p,n}(M, E) \xrightarrow{\overline{\partial}^{p,n}} 0. \tag{2.93}$$

The cohomology $\mathrm{Ker}(\overline{\partial}^{p,q})/\mathrm{Im}(\overline{\partial}^{p,q-1})$ of this complex is called the Dolbeault cohomology of a holomorphic vector bundle of E and is denoted by $H^{p,q}_{\overline{\partial}}(M, E)$. Unlike the case of $H^{p,q}_{\overline{\partial}}(M)$, this cohomology is not related to the De Rham cohomology of M (it is related to the Čech cohomology of sheaves which take values in $E \otimes (\bigwedge^p T'M)$). But in the $q = 0$ case, it has a specific geometrical interpretation. In this case, the condition that the form given in (2.92) is closed reduces to

$$\overline{\partial}(f^E_{i_1 i_2 \cdots i_p} dz^{i_1} \wedge \cdots \wedge dz^{i_p}) = 0, \Longleftrightarrow \partial_{\bar{j}} f^E_{i_1 i_2 \cdots i_p} \overline{dz^j} \wedge dz^{i_1} \wedge \cdots \wedge dz^{i_p} = 0, \Longleftrightarrow \partial_{\bar{j}} f^E_{i_1 i_2 \cdots i_p} = 0, \ (j = 1, 2, \ldots, n), \tag{2.94}$$

and it is equivalent to the condition that all the components of $f^E_{i_1 i_2 \cdots i_p}$ are holomorphic. From this, we can conclude that $\dim(H^{p,0}_{\overline{\partial}}(M, E))$ is the number of linearly independent global holomorphic sections of $E \otimes (\wedge^p T'M)$. Therefore, the Hodge number $h^{p,0}(M) = \dim(H^{p,0}_{\overline{\partial}}(M))$ is nothing but the number of linearly independent holomorphic sections of $\wedge^p T'M$, or the number of linearly independent global holomorphic p-forms on M.

2.5 Complex Projective Space as an Example of Compact Kähler Manifold

In this section, we discuss the n-dimensional complex projective space CP^n and the degree k hypersurface in CP^n as important examples of compact Kähler manifolds. In order to show that CP^n is a Kähler manifold, we have to construct the Kähler metric $g_{i\bar{j}}$ on CP^n. We construct $g_{i\bar{j}}$ by using a complex line bundle on CP^n. The line bundle we use is called a tautological line bundle on CP^n and it is denoted by S. S is a line bundle whose fiber on $(X_1 : X_2 : \cdots : X_{n+1}) \in CP^n$ is given by a complex 1-dimensional vector space in $\mathbf{C}^{n+1}$ spanned by $(X_1, X_2, \ldots, X_{n+1})$. Let $U_i \subset CP^n$ be $\{ (X_1 : X_2 : \cdots : X_{n+1}) \in CP^n \mid X_i \neq 0 \}$. Then we can take the local complex coordinate system $\phi_{(i)} : U_i \to \mathbf{C}^n$:

$$\phi_{(i)}(X_1; X_2 : \cdots : X_{n+1}) = \left(\frac{X_1}{X_i}, \ldots, \frac{X_{i-1}}{X_i}, \frac{X_{i+1}}{X_i}, \ldots, \frac{X_{n+1}}{X_i}\right) = (z^1_{(i)}, z^2_{(i)}, \ldots, z^n_{(i)}).$$

We take $CP^n = \bigcup_{i=1}^{n+1} U_i$ as an open covering used for local trivialization of S and determine $e^{(i)}$, the base of fiber vector space for trivialization of S with respect to U_i as follows:

$$e^{(i)} := (z^1_{(i)}, z^2_{(i)}, \ldots, z^{i-1}_{(i)}, 1, z^i_{(i)}, \ldots, z^n_{(i)}) \in \mathbf{C}^{n+1}. \tag{2.95}$$

If $j < i$, $e^{(i)}$ is represented by $e^{(j)}$ on $U_i \cap U_j$ as follows:

$$\begin{aligned} e^{(i)} &= (z^1_{(i)}, z^2_{(i)}, \ldots, z^{i-1}_{(i)}, 1, z^i_{(i)}, \ldots, z^n_{(i)}) \\ &= z^j_{(i)}(z^1_{(j)}, \ldots, z^{j-1}_{(j)}, 1, z^j_{(j)}, \ldots, z^n_{(j)}) = e^{(j)} z^j_{(i)}. \end{aligned} \tag{2.96}$$

Hence the transition function g_{ji} is given by $z^j_{(i)}$ and it is holomorphic. (If $j > i$, $e^{(i)} = e^{(j)} z^{j-1}_{(i)}$.) In this way, we can see that S is a holomorphic line bundle. We next construct a Hermitian metric of S. We use the natural metric $\sum_{j=1}^{n+1} x^j \overline{x^j}$ of $\mathbf{C}^{n+1} = \{ (x^1, x^2, \ldots, x^{n+1}) \}$ and define it by

$$h^{(i)} = \langle e^{(i)}, e^{(i)} \rangle = 1 + \sum_{j=1}^{n} z^j_{(i)} \overline{z^j_{(i)}}. \tag{2.97}$$

From the discussion used in the proof of Theorem 2.3.4, the curvature $R^{(i)}_{k\bar{l}}$ compatible with the metric and the differential form that corresponds to $c_1(R^{(i)})$ are obtained as follows:

$$R^{(i)}_{k\bar{l}} = -\partial_{\bar{l}}\partial_k \log(1+\sum_{j=1}^{n} z^j_{(i)}\overline{z^j_{(i)}}),\ c_1(R^{(i)}) = -\frac{i}{2\pi}\partial\bar{\partial}\log(1+\sum_{j=1}^{n} z^j_{(i)}\overline{z^j_{(i)}}). \tag{2.98}$$

Since $c_1(R^{(i)})$ defines a global (1, 1) form on CP^n, $R^{(i)}_{k\bar{l}}$ behaves in the same way as a Hermitian metric of $T'CP^n$ on $U_i \cap U_j$. Therefore, we can set $g^{(i)}_{k\bar{l}} = -R^{(i)}_{k\bar{l}} = \partial_{\bar{l}}\partial_k \log(1+\sum_{j=1}^{n} z^j_{(i)}\overline{z^j_{(i)}})$ as a positive definite Hermitian metric of $T'CP^n$. Moreover, since $g = g^{(i)}_{k\bar{l}} dz^k_{(i)} \wedge d\overline{z^l_{(i)}} = \partial\bar{\partial}\log(1+\sum_{j=1}^{n} z^j_{(i)}\overline{z^j_{(i)}})$, the condition $dg = (\partial+\bar{\partial})g = 0$ holds. In sum, we have shown the following theorem.

Theorem 2.5.1 *The Hermitian metric of $T'CP^n$ defined by*

$$g^{(i)}_{k\bar{l}} := \partial_{\bar{l}}\partial_k \log(1+\sum_{j=1}^{n} z^j_{(i)}\overline{z^j_{(i)}}) \tag{2.99}$$

satisfies the condition that $g = g^{(i)}_{k\bar{l}} dz^k_{(i)} \wedge d\overline{z^l_{(i)}}$ is closed. Hence CP^n with this metric is a Kähler manifold.

In the above construction, the tautological line bundle S on CP^n played an important role. Now, let us consider the dual line bundle S^*. Let $e^{*(i)}$ be the base used for local trivialization of S^* in U_i. In $U_i \cap U_j$, the relation:

$$e^{*(i)} = e^{*(j)}\frac{1}{z^j_{(i)}},\ (j < i),\ e^{*(i)} = e^{*(j)}\frac{1}{z^{j-1}_{(i)}},\ (j > i), \tag{2.100}$$

follows from (2.96). Now, let us consider the following tuple of holomorphic functions on U_i $(i = 1, 2, \dots, n+1)$,

$$f^m_{(i)} = z^m_{(i)},\ (m < i),\ f^m_{(i)} = 1,\ (m = i),\ f^m_{(i)} = z^{m-1}_{(i)},\ (i < m), \tag{2.101}$$

and compare $f^m_{(i)}e^{*(i)}$ with $f^m_{(j)}e^{*(j)}$ in $U_i \cap U_j$. We have to consider various cases with respect to the magnitude relation of m, i and j, and consider the case when $m < i < j$ for example. We can easily see from (2.100) and (2.14) that they coincide:

$$f^m_{(i)}e^{*(i)} = z^m_{(i)}e^{*(j)}\frac{1}{z^j_{(i)}} = \frac{z^m_{(i)}}{z^j_{(i)}}e^{*(j)} = z^m_{(j)}e^{*(j)} = f^m_{(j)}e^{*(j)}. \tag{2.102}$$

By similar computation, we can confirm that $f^m_{(i)}e^{*(i)} = f^m_{(j)}e^{*(j)}$ holds in the other cases. This fact asserts that we can patch $f^m_{(i)}e^{*(i)}$ $(i = 1, 2, \dots, n+1)$ and define global holomorphic sections s^m $(m = 1, 2, \dots, n+1)$ of S^*. On the other hand, S does not admit such global holomorphic sections. In this sense, S^* plays more important roles than S in the geometry of CP^n. An example of this importance is given by the close relation of S^* and homogeneous coordinates X_i, $(i = 1, 2, \ldots, n)$ of CP^n. In $U_i \cap U_j$, we can easily derive

$$X_i = X_j \frac{1}{z_{(i)}^j} \ (j < i), \quad X_i = X_j \frac{1}{z_{(i)}^{j-1}} \ (i < j), \tag{2.103}$$

from $X_i = \dfrac{X_i}{X_j} X_j$. (2.103) coincides with the transition rule (2.100) of $e^{*(i)}$. Hence we can formally identify $e^{*(i)}$ with X_i. We then focus on the global holomorphic section s^m. Since $f_{(i)}^m$ is rewritten as $\dfrac{X_m}{X_i}$ by using homogeneous coordinates, we obtain

$$f_{(i)}^m e^{*(i)} = \frac{X_m}{X_i} X_i = X_m. \tag{2.104}$$

Therefore, we can identify s^m with the homogeneous coordinate X_m. The fact that global holomorphic sections of S^* are generated by X_m $(m = 1, 2, \ldots, n + 1)$ is the origin of the well-known naming of S^*, "hyperplane bundle H" (this is because $X_m = 0$ defines a hyperplane in CP^n which is biholomorphic to CP^{n-1}).

At this stage, let us look back at the degree k hypersurface M_n^k in CP^n, which was introduced in Sect. 2.1. The degree k homogeneous polynomial in X_m's used to define M_n^k is nothing but a global holomorphic section of $H^{\otimes k}$. Hence M_n^k is characterized as the vanishing locus of a global holomorphic section of $H^{\otimes k}$. In algebraic geometry, $H^{\otimes k}$ is usually represented by $\mathcal{O}_{CP^n}(k)$. From now on, we also use the notation $\mathcal{O}_{CP^n}(k)$ in this book. By pulling back the Kähler metric of CP^n in Theorem 2.5.1 via inclusion map $i : M_n^k \hookrightarrow CP^n$, we can construct an Hermitian metric on M_n^k. The associated (1, 1) form is given by $i^*(g)$ and it is a closed form. Hence M_n^k is also a Kähler manifold.

In order to give concrete examples of Chern classes, we evaluate $c(CP^n) = c(T'CP^n)$ and $c(M_n^k) = c(T'M_n^k)$. To compute $c(T'CP^n)$, we use the fact that $T'CP^n$ can be represented by S and $S^* = H$. Note that S is the sub-vector bundle of a trivial bundle $CP^n \times \mathbf{C}^{n+1}$. Let $S_p \subset \mathbf{C}^{n+1}$ be the fiber vector space of S at $p \in CP^n$ and Q_p be a quotient vector space $\mathbf{C}^{n+1}/S_p$. We can summarize this situation into a short exact sequence of vector spaces:

$$0 \to S_p \overset{i_p}{\to} \{p\} \times \mathbf{C}^{n+1} \overset{\pi_p}{\to} Q_p \to 0. \tag{2.105}$$

Let Q be a vector bundle on CP^n whose fiber at $p \in CP^n$ is given by Q_p. The above exact sequence turns into an exact sequence of vector bundles:

$$0 \to S \overset{i}{\to} \mathbf{C}^{n+1} \overset{\pi}{\to} Q \to 0. \tag{2.106}$$

Then let us consider T'_pCP^n, i.e., the holomorphic tangent space of CP^n at p. Since $p \in CP^n$ can be regarded as a 1-dimensional complex vector space S_p in $\mathbf{C}^{n+1}$, a tangent vector at p, or infinitesimal translation of S_p, is given by a linear map from S_p to Q_p. Therefore, T'_pCP^n is identified with $\mathrm{Hom}_{\mathbf{C}}(S_p, Q_p) = S_p^* \otimes Q_p = H_p \otimes Q_p$. In this way, we can conclude that $T'CP^n$ is isomorphic to $H \otimes Q$ as a vector bundle. Hence we obtain the following exact sequence of vector bundles by taking the tensor product of (2.106) with H,

$$0 \to \mathbf{C} \to H^{\oplus n+1} \to T'CP^n \to 0, \tag{2.107}$$

where we used the relation $S \otimes H = S \otimes S^* = \mathbf{C}$ (we denote here the trivial vector bundle $CP^n \times \mathbf{C}$ by $\mathbf{C}$). From this exact sequence, we can effectively regard the rank $n+1$ vector bundle $H^{\oplus n+1}$ as a direct sum $\mathbf{C} \oplus T'CP^n$ and derive the following equality by using Theorem 2.3.3:

$$c(T'CP^n) = c(H^{\oplus n+1}) = (c(H))^{n+1} = (1 + tc_1(H))^{n+1}, \tag{2.108}$$

where we used the fact $c(\mathbf{C}) = 1$. From now on, we denote $c_1(H) \in H^{1,1}(CP^n)$ by h. It is well-known that the de Rham cohomology ring of CP^n is isomorphic to a polynomial ring generated by h with one relation $h^{n+1} = 0$. Hence non-trivial elements of $H^{*,*}(CP^n, \mathbf{C})$ are given by $h^j \in H^{j,j}(CP^n)$ $(j = 0, 1, \ldots, n)$. These facts and (2.108) lead us to the formula:

$$c(CP^n) = c(T'CP^n) = \sum_{j=0}^{n} t^j \binom{n+1}{j} h^j. \tag{2.109}$$

Especially, $c_1(CP^n)$ is given by $(n+1)h$. Moreover, $h^n \in H^{n,n}(CP^n, \mathbf{C})$ satisfy the following normalization condition:

$$\int_{CP^n} h^n = 1. \tag{2.110}$$

Therefore, we can see from Theorem (2.3.5) and (2.110) that $\chi(CP^n)$ equals $n+1$.

Next, we evaluate $c(M_n^k) = c(T'M_n^k)$. Let $i : M_n^k \hookrightarrow CP^n$ be the inclusion map. Then we can regard $T'M_n^k$ as a sub-vector bundle of $i^{-1}T'CP^n$. The quotient vector bundle $i^{-1}T'CP^n/T'M_n^k$ is a normal vector bundle of M_n^k in CP^n and it is isomorphic to $i^{-1}\mathcal{O}_{CP^n}(k)$, where $H^{\otimes k} = \mathcal{O}_{CP^n}(k)$ is the line bundle whose holomorphic section is nothing but the defining equation of M_n^k. Hence we obtain the following short exact sequence of vector bundles on M_N^k:

$$0 \to T'M_n^k \to i^{-1}T'CP^n \to i^{-1}\mathcal{O}_{CP^n}(k) \to 0. \tag{2.111}$$

This leads us to an equality $c(T'M_n^k)i^*c(\mathcal{O}_{CP^n}(k)) = i^*c(T'CP^n)$. Combining $i^*c(\mathcal{O}_{CP^n}(k)) = 1 + t(i^*c_1(\mathcal{O}_{CP^n}(k))) = 1 + t(i^*c_1(H^{\otimes k})) = 1 + t(ki^*h)$ with (2.109), we obtain

$$c(M_n^k) = \sum_{j=0}^{n-1} t^j c_j(M_n^k) = \left(\sum_{j=0}^{n} t^j \binom{n+1}{j} (i^*h)^j\right) / (1 + tk(i^*h)). \quad (2.112)$$

In the above equality, we expand formally the r.h.s in t and take terms up to $(n-1)$-th power of t. From this formula, we can see that $c_1(M_n^k)$ equals $(n+1-k)i^*h$ and that it vanishes if $n+1=k$. Therefore, M_n^{n+1} is a Calabi–Yau manifold. M_4^5 is the complex 3-dimensional Calabi–Yau manifold which Candelas et al. used in their renowned paper on mirror symmetry. The explicit form of $c(M_4^5)$ is easily obtained by using (2.112) as follows:

$$c(M_4^5) = 1 + t^2(10(i^*h)^2) + t^3(-40(i^*h)^3). \quad (2.113)$$

Lastly, we evaluate $\chi(M_4^5)$ by using Theorem 2.3.5. From the theorem, the Euler number is given by

$$\chi(M_4^5) = \int_{M_4^5} (-40(i^*h)^3). \quad (2.114)$$

Here, we use a well-known formula that translates integration of a degree (3, 3) cohomology element on M_4^5 into integration on CP^4 via the inclusion map $i : M_4^5 \hookrightarrow CP^4$:

$$\int_{M_4^5} i^*\alpha = \int_{CP^4} c_1(\mathcal{O}_{CP^4}(5)) \cdot \alpha = \int_{CP^4} (5h) \cdot \alpha. \quad (2.115)$$

This leads us to the result:

$$\chi(M_4^5) = \int_{CP^4} (5h) \cdot (-40h^3) = -200. \quad (2.116)$$

Reference

1. P. Griffiths, J. Harris. *Principles of Algebraic Geometry.* Wiley–Interscience (1978).

Chapter 3
Topological Sigma Models

Abstract In this chapter, we discuss topological sigma models, which enable us to interpret Yukawa couplings used in mirror symmetry as correlation functions of topological field theory. There are two types of topological sigma models, both of which are obtained from the $N = 2$ supersymmetric sigma model through an operation called "topological twist". This distinction comes from the differences in the topological twist. We call these two types A-model and B-model respectively [1]. We first introduce the $N = 2$ supersymmetric sigma model and then construct the Lagrangian of an the A-model via the A-twist of the supersymmetric sigma model. Next, we discuss correlation function of the A-model explains that Yukawa coupling associated with Kähler deformation of a Calabi–Yau 3-fold M can be regarded as a three-point correlation function of the A-model on M. Lastly, we construct the Lagrangian of the B-model via the B-twist of the supersymmetric sigma model and explain that Yukawa coupling associated with complex structure deformation of a Calabi–Yau 3-fold M^* can be regarded as a three-point correlation function of the B-model on M^*.

3.1 $N = 2$ Supersymmetric Sigma Model

In this section, we introduce the $N = 2$ supersymmetric sigma model. It is a model that describes dynamics of maps from a 2-dimensional Riemann surface Σ to a Kähler manifold M. It also has $N = 2$ supersymmetry. Therefore, its dynamical variables consist of bosonic fields, $\phi^i(z, \overline{z})$ and their complex conjugate $\phi^{\overline{i}}(z, \overline{z}) = \overline{\phi^i(z, \overline{z})}$, which describe maps from Σ to M, and fermionic fields, $\psi^i_+(z, \overline{z})$,$\psi^i_-(z, \overline{z})$, $\psi^{\overline{i}}_+(z, \overline{z})$ and $\psi^{\overline{i}}_-(z, \overline{z})$, which are superpartners of bosonic fields. We use the subscripts i and $\overline{i}$ to distinguish local complex coordinates of M and its complex conjugate. The subscripts $+$ and $-$ correspond to spin quantum numbers $\frac{1}{2}$ and $-\frac{1}{2}$ in 2-dimensional spinor representation. With these fields, the Lagrangian L is given by

M. Jinzenji, *Classical Mirror Symmetry*, SpringerBriefs in Mathematical Physics,
https://doi.org/10.1007/978-981-13-0056-1_3

$$L = 2t \int_\Sigma d^2z \Big(\frac{1}{2} g_{i\bar{j}} (\partial_z \phi^i \partial_{\bar{z}} \phi^{\bar{j}} + \partial_z \phi^{\bar{j}} \partial_{\bar{z}} \phi^i)$$
$$+ i g_{\bar{i}j} \psi_-^{\bar{j}} D_z \psi_-^i + i g_{\bar{i}j} \psi_+^{\bar{j}} D_{\bar{z}} \psi_+^i + R_{i\bar{j}k\bar{l}} \psi_+^i \psi_+^{\bar{j}} \psi_-^k \psi_-^{\bar{l}} \Big). \tag{3.1}$$

In (3.1), covariant derivatives D_z and $D_{\bar{z}}$ for fermions are defined by

$$D_z \psi_-^i := \frac{\partial}{\partial z} \psi_-^i + \frac{\partial \phi^j}{\partial z} \Gamma^i_{jk} \psi_-^k, \quad D_{\bar{z}} \psi_+^i := \frac{\partial}{\partial \bar{z}} \psi_+^i + \frac{\partial \phi^j}{\partial \bar{z}} \Gamma^i_{jk} \psi_+^k. \tag{3.2}$$

Here, Γ^i_{jk} is the connection of $T'M$ defined in Sect. 2.3.1. The tensor $R_{i\bar{j}k\bar{l}}$ is obtained from the curvature tensor $R^i_{\ jk\bar{l}}$ of $T'M$ via relations $R_{\bar{j}ik\bar{l}} = g_{m\bar{j}} R^m_{\ ik\bar{l}}$ and $R_{i\bar{j}k\bar{l}} = -R_{\bar{j}ik\bar{l}}$. This Lagrangian is invariant under the supersymmetric transformation

$$\begin{aligned}
&\delta\phi^i = i\alpha_- \psi_+^i + i\alpha_+ \psi_-^i, \ \delta\phi^{\bar{i}} = i\tilde{\alpha}_- \psi_+^{\bar{i}} + i\tilde{\alpha}_+ \psi_-^{\bar{i}}, \\
&\delta\psi_+^i = -\tilde{\alpha}_- \partial_z \phi^i - i\alpha_+ \psi_-^j \Gamma^i_{jm} \psi_+^m, \ \delta\psi_+^{\bar{i}} = -\alpha_- \partial_z \phi^{\bar{i}} - i\tilde{\alpha}_+ \psi_-^{\bar{j}} \Gamma^{\bar{i}}_{\bar{j}\bar{m}} \psi_+^{\bar{m}}, \\
&\delta\psi_-^i = -\tilde{\alpha}_+ \partial_{\bar{z}} \phi^i - i\alpha_- \psi_+^j \Gamma^i_{jm} \psi_-^m, \ \delta\psi_-^{\bar{i}} = -\alpha_+ \partial_{\bar{z}} \phi^{\bar{i}} - i\tilde{\alpha}_- \psi_+^{\bar{j}} \Gamma^{\bar{i}}_{\bar{j}\bar{m}} \psi_-^{\bar{m}}.
\end{aligned} \tag{3.3}$$

The four Grassmann parameters $\alpha_+, \alpha_-, \tilde{\alpha}_+, \tilde{\alpha}_-$ are independent and the subscripts $+$ and $-$ correspond to spin quantum numbers $\frac{1}{2}$ and $-\frac{1}{2}$ respectively. $\Gamma^{\bar{i}}_{\bar{j}\bar{k}}$ is the connection of the anti-holomorphic tangent vector bundle $\overline{T'M}$ and satisfies $\Gamma^{\bar{i}}_{\bar{j}\bar{k}} = \overline{\Gamma^i_{jk}}$. These four independent supersymmetries imply that the theory has two independent supersymmetries with respect to the right-handed (holomorphic in z) degree of freedom and the left-handed (anti-holomorphic in z) degree of freedom. Therefore, this supersymmetry is called $N = (2, 2)$ supersymmetry. It is known that a theory with $N = (2, 2)$ supersymmetry automatically has $U(1)$ symmetry. The $U(1)$ charges of $\phi^i, \phi^{\bar{i}}$ are 0, those of ψ_+^i, ψ_-^i are 1 and those of $\psi_+^{\bar{i}}, \psi_-^{\bar{i}}$ are -1 respectively. The topological twist for this theory is the operation to add (or subtract) half of the $U(1)$ charges of the fields to (or from) the spin quantum numbers of the fields. The operation to subtract half of the $U(1)$ charges from both the field with $+$ spin and the field with $-$ spin is called the A-twist. On the other hand, the operation to add half of the $U(1)$ charge to the field with $+$ spin and to subtract it from the field with $-$ spin is called the B-twist. Therefore, in the case of A-twist, fields change in the following way:

$$\phi^i, \phi^{\bar{i}} \to \phi^i, \phi^{\bar{i}}, \quad \psi_+^i \to \psi^i, \ \psi_-^i \to \psi_{\bar{z}}^i, \ \psi_+^{\bar{i}} \to \psi_z^{\bar{i}}, \ \psi_-^{\bar{i}} \to \psi^{\bar{i}}. \tag{3.4}$$

In the case of B-twist, the change of fields is given by

$$\phi^i, \phi^{\bar{i}} \to \phi^i, \phi^{\bar{i}}, \quad \psi_+^i \to \psi_z^i, \ \psi_-^i \to \psi_{\bar{z}}^i, \ \psi_+^{\bar{i}} \to \psi^{\bar{i}}, \ \psi_-^{\bar{i}} \to \psi^{\bar{i}}. \tag{3.5}$$

(In the case of B-twist, fields obtained from $\psi_+^{\bar{i}}$ and $\psi_-^{\bar{i}}$ have the same spin and cannot be distinguished from each other by notation. But we consider them different fields. We will discuss this situation later in this chapter).

We roughly explain how these twists affect the $N = (2, 2)$ supersymmetric sigma model. From a physical point of view, the sigma model describes the dynamics of a real 1-dimensional string moving in a Kähler manifold M, and its quantum Hilbert space consists of zero modes that correspond to the solution of an instanton equation and oscillation modes with positive energy eigenvalues. In the usual field theory, analysis of oscillation modes is difficult and important, but the "twist" operation makes bosonic oscillation modes and fermionic oscillation modes cancel each other. Hence the non-trivial part of the theory obtained after twist turns out to be structure of zero modes, or the solution space of the instanton equation. Let us call this space the moduli space of instantons. In the next section, we will show that the correlation function of the twisted theory is identified with the topological intersection number the of moduli space of instantons. Therefore, the twisted $N = (2, 2)$ supersymmetric sigma model is called the topological sigma model. In the case of A-twist, the instanton is given by a holomorphic map from Σ to M, which is called the "world sheet instanton" in the context of string theory. Instanton correction of the Yukawa coupling associated with Kähler deformation of the Calabi–Yau 3-fold M comes from this world sheet instanton. In the case of B-twist, the instanton turns out to be a constant map from Σ to M. This fact is deeply connected with the fact that Yukawa coupling associated with complex structure deformation of the Calabi–Yau 3-fold M is represented by the classical integral of differential form on M. Therefore, we can regard the classical mirror symmetry conjecture as a coincidence between the correlation function of the A-twisted model on the Calabi–Yau 3-fold X and the correlation function of B-twisted model on the mirror partner Calabi–Yau 3-fold X^*.

3.2 Topological Sigma Model (A-Model)

3.2.1 *Lagrangian and Saddle Point Approximation*

As we have discussed in the previous section, the topological sigma model (A-model) is obtained from the $N = (2, 2)$ supersymmetric sigma model by subtracting half of the $U(1)$ charges of fields from their spin quantum numbers. Therefore, fields that appear in the A-model are given by

$$\phi^i, \phi^{\bar{i}} \to \phi^i, \phi^{\bar{i}}, \quad \psi_+^i \to \psi^i, \ \psi_-^i \to \psi_{\bar{z}}^i, \ \psi_+^{\bar{i}} \to \psi_z^{\bar{i}}, \ \psi_-^{\bar{i}} \to \psi^{\bar{i}}. \tag{3.6}$$

Let us denote the fermions ψ^i and $\psi^{\bar{i}}$ by χ^i and $\chi^{\bar{i}}$ respectively. Then the Lagrangian of the A-model is given as follows:

$$L = 2t \int_\Sigma d^2z \Big(\frac{1}{2} g_{i\bar{j}} (\partial_z \phi^i \partial_{\bar{z}} \phi^{\bar{j}} + \partial_z \phi^{\bar{j}} \partial_{\bar{z}} \phi^i) + i g_{\bar{i}j} \psi_z^{\bar{j}} D_{\bar{z}} \chi^i + i g_{i\bar{j}} \psi_{\bar{z}}^i D_z \chi^{\bar{j}} - R_{i\bar{j}k\bar{l}} \psi_{\bar{z}}^i \psi_z^{\bar{j}} \chi^k \chi^{\bar{l}} \Big). \tag{3.7}$$

The supersymmetric transformation of this theory is obtained by setting $\alpha_- = \tilde{\alpha}_+ = 0$ and $\alpha_+ = \tilde{\alpha}_- = \alpha$ in (3.3):

$$\begin{aligned} &\delta\phi^i = i\alpha\chi^i, \quad \delta\phi^{\bar{i}} = i\tilde{\alpha}\chi^{\bar{i}}, \\ &\delta\chi^i = 0, \quad \delta\chi^{\bar{i}} = 0, \\ &\delta\psi_z^{\bar{i}} = -\alpha\partial_z\phi^{\bar{i}} - i\tilde{\alpha}\chi^{\bar{j}}\Gamma^{\bar{i}}_{\bar{j}\bar{m}}\psi_z^{\bar{m}}, \quad \delta\psi_{\bar{z}}^i = -\tilde{\alpha}\partial_{\bar{z}}\phi^i - i\alpha\chi^j\Gamma^i_{jm}\psi_{\bar{z}}^m. \end{aligned} \tag{3.8}$$

Let us define the generator Q of this transformation via the relation $\delta X = -i\alpha\{Q, X\}$ (X is a field that appears in the theory). Then we can easily see that $\{Q, \{Q, X\}\} = 0$ holds. If we rewrite the relation to $Q^2X = 0$, we can regard the operator Q as an analogue of the exterior differential operator d in the theory of differential forms on manifolds. In field theory, the fermionic transformation whose generator Q satisfies $Q^2X = 0$ is called the "BRST-transformation". Of course, the Lagrangian L in (3.7) satisfies $QL = 0$. Hence this model has BRST-symmetry.

By making use of the ψ-equation of motion,

$$i\psi_z^{\bar{j}} D_{\bar{z}} \chi^i g_{i\bar{j}} = R_{i\bar{j}k\bar{l}} \psi_{\bar{z}}^i \psi_z^{\bar{j}} \chi^k \chi^{\bar{l}}, \quad i\psi_{\bar{z}}^i D_z \chi^{\bar{j}} g_{i\bar{j}} = R_{i\bar{j}k\bar{l}} \psi_{\bar{z}}^i \psi_z^{\bar{j}} \chi^k \chi^{\bar{l}}, \tag{3.9}$$

we can rewrite L into the following form:

$$L = it \int_\Sigma d^2z \{Q, V\} + t \int_\Sigma \Phi^*(K). \tag{3.10}$$

In (3.10), V is given by

$$V = g_{i\bar{j}} (\psi_z^{\bar{j}} \partial_{\bar{z}} \phi^i + \psi_{\bar{z}}^i \partial_z \phi^{\bar{j}}), \tag{3.11}$$

and the explicit form of the second term is

$$\int_\Sigma \Phi^*(K) = \int_\Sigma d^2z \big(g_{i\bar{j}} \partial_z \phi^i \partial_{\bar{z}} \phi^{\bar{j}} - g_{i\bar{j}} \partial_{\bar{z}} \phi^i \partial_z \phi^{\bar{j}} \big). \tag{3.12}$$

This second term is the pull-back of the Kähler form $K = g_{i\bar{j}} dz^i \wedge dz^{\bar{j}}$ of M by the map $\phi : \Sigma \to M$. Since $K = g_{i\bar{j}} dz^i \wedge dz^{\bar{j}}$ is a representative of $H^2(M)$, $\int_\Sigma \Phi^*(K)$ becomes a topological invariant. Especially if $H^2(M, \mathbf{Z}) \simeq \mathbf{Z}$, it becomes

$$\int_\Sigma \Phi^*(K) = 2\pi i d \quad (d \text{ is a non-negative integer}), \tag{3.13}$$

by suitably adjusting the scale of $g_{i\bar{j}}$. This d is called the winding number, or degree of the map ϕ, and it is a homotopy invariant (the homotopy invariant is a quantity invariant under continuous deformation of the map).

Now, we discuss general characteristics of the A-model derived from the fact that it is invariant under BRST-transformation. First, we represent the correlation function of observable W constructed from fields in the following way:

$$\langle W \rangle := \int \mathscr{D}X e^{-L} W. \tag{3.14}$$

If W is written as $W = \{Q, U\}$, where U is some other observable, the corresponding correlation function vanishes:

$$\langle \{Q, U\} \rangle := \int \mathscr{D}X e^{-L} \{Q, U\} = 0 \tag{3.15}$$

This can be shown as follows. Let us consider "rotating" the field by using supersymmetric transformation. This rotation is realized by acting the operator $\exp(\varepsilon Q)$ on fields where ε is a Grassmann number. Since the path-integral is invariant under rotation of fields, the following relation holds:

$$\begin{aligned} \langle U \rangle &= \int \mathscr{D}X e^{-L} U = \int \mathscr{D}(\exp(\varepsilon Q) X)(\exp(\varepsilon Q) e^{-L} U) \\ &= \int \mathscr{D}X \,(\exp(\varepsilon Q) e^{-L} U). \end{aligned} \tag{3.16}$$

In the above derivation, we used the relation $\mathscr{D}(\exp(\varepsilon Q) X) = \mathscr{D}X$ that follows from the fact that the Jacobian induced from the coordinate change of rotation is 1. The last line of (3.16) is further rewritten by using the relation $\{Q, L\} = 0$ and the fact that ε is a Grassmann number:

$$\begin{aligned} \langle U \rangle &= \int \mathscr{D}X \,(\exp(\varepsilon Q) e^{-L} U) = \int \mathscr{D}X e^{-L} (\exp(\varepsilon Q) U) \\ &= \int \mathscr{D}X e^{-L} (U + \varepsilon \{Q, U\}) = \langle U \rangle + \varepsilon \langle \{Q, U\} \rangle, \end{aligned} \tag{3.17}$$

Since ε is arbitrary, $\langle \{Q, U\} \rangle = 0$ holds.

In studying a theory with BRST-symmetry, we restrict the observable $\mathscr{O}$ to satisfy the relation $\{Q, \mathscr{O}\} = 0$. One of the reasons for this restriction comes from the following well-known formula.

"If observables $\mathscr{O}_i$ $(i = 1, \ldots, n)$ satisfy $\{Q, \mathscr{O}_i\} = 0$, the equalities,

$$\{Q, \mathscr{O}_1 \mathscr{O}_2 \cdots \mathscr{O}_n\} = 0, \quad \{Q, A\} \mathscr{O}_1 \mathscr{O}_2 \cdots \mathscr{O}_n = \{Q, A \mathscr{O}_1 \mathscr{O}_2 \cdots \mathscr{O}_n\} \tag{3.18}$$

hold for arbitrary observable A."

This formula is obvious since the equality:

$$\{Q, A_1 A_2 \cdots A_n\} = \sum_{i=1}^{n} (\text{sign.}) A_1 \cdots A_{i-1} \{Q, A_i\} A_{i+1} \cdots A_n, \tag{3.19}$$

holds for arbitrary observables A_i $(i = 1, 2, \ldots, n)$. From now on, we assume that the observable $\mathcal{O}$ satisfies the relation $\{Q, \mathcal{O}\} = 0$. Then let us consider the correlation function $\langle \mathcal{O}_1 \mathcal{O}_2 \cdots \mathcal{O}_n \rangle$. By using the homotopy invariant d of the map ϕ, we can decompose the phase space of the path-integral into connected components P_d $(d = 0, 1, 2, \cdots)$. By combining this decomposition with (3.10), the correlation function is rewritten as follows:

$$\begin{aligned}
\langle \mathcal{O}_1 \mathcal{O}_2 \cdots \mathcal{O}_n \rangle &= \int \mathcal{D}X e^{-L} \mathcal{O}_1 \mathcal{O}_2 \cdots \mathcal{O}_n \\
&= \int \mathcal{D}X e^{-it \int_\Sigma d^2z \{Q, V\} - t \int_\Sigma \Phi^*(K)} \mathcal{O}_1 \mathcal{O}_2 \cdots \mathcal{O}_n \\
&= \sum_{d=0}^{\infty} e^{-2\pi i d t} \int_{P_d} \mathcal{D}X e^{-it \int_\Sigma d^2z \{Q, V\}} \mathcal{O}_1 \mathcal{O}_2 \cdots \mathcal{O}_n.
\end{aligned} \tag{3.20}$$

If we set

$$\langle \mathcal{O}_1 \mathcal{O}_2 \cdots \mathcal{O}_n \rangle_d = \int_{P_d} \mathcal{D}X e^{-it \int_\Sigma d^2z \{Q, V\}} \mathcal{O}_1 \mathcal{O}_2 \cdots \mathcal{O}_n, \tag{3.21}$$

the correlation function is represented in the form:

$$\langle \mathcal{O}_1 \mathcal{O}_2 \cdots \mathcal{O}_n \rangle = \sum_{d=0}^{\infty} e^{-2\pi i d t} \langle \mathcal{O}_1 \mathcal{O}_2 \cdots \mathcal{O}_n \rangle_d. \tag{3.22}$$

We call $\langle \mathcal{O}_1 \mathcal{O}_2 \cdots \mathcal{O}_n \rangle_d$ the degree d correlation function. If we apply the previous discussion to the r.h.s. of (3.21), we can show that the degree d correlation function does not depend on the coupling constant t. Let us vary t into $t + \delta t$ in the r.h.s. of (3.21) (we assume that δt is infinitesimally small):

$$\begin{aligned}
&\int_{P_d} \mathcal{D}X e^{-i(t+\delta t) \int_\Sigma d^2z \{Q, V\}} \mathcal{O}_1 \mathcal{O}_2 \cdots \mathcal{O}_n \\
&= \int_{P_d} \mathcal{D}X e^{-it \int_\Sigma d^2z \{Q, V\}} e^{-i\delta t \int_\Sigma d^2z \{Q, V\}} \mathcal{O}_1 \mathcal{O}_2 \cdots \mathcal{O}_n \\
&= \int_{P_d} \mathcal{D}X e^{-it \int_\Sigma d^2z \{Q, V\}} (1 - i\delta t \int_\Sigma d^2z \{Q, V\}) \mathcal{O}_1 \mathcal{O}_2 \cdots \mathcal{O}_n \\
&= \int_{P_d} \mathcal{D}X e^{-it \int_\Sigma d^2z \{Q, V\}} \mathcal{O}_1 \mathcal{O}_2 \cdots \mathcal{O}_n
\end{aligned}$$

$$
\begin{aligned}
&- i\delta t \int_{P_d} \mathscr{D}X e^{-it\int_\Sigma d^2z\{Q,V\}}\{Q, \int_\Sigma d^2zV\}\mathscr{O}_1\mathscr{O}_2\cdots\mathscr{O}_n \\
&= \int_{P_d} \mathscr{D}X e^{-it\int_\Sigma d^2z\{Q,V\}}\mathscr{O}_1\mathscr{O}_2\cdots\mathscr{O}_n \\
&- i\delta t \int_{P_d} \mathscr{D}X e^{-it\int_\Sigma d^2z\{Q,V\}}\{Q, (\int_\Sigma d^2zV)\mathscr{O}_1\mathscr{O}_2\cdots\mathscr{O}_n\} \\
&= \int_{P_d} \mathscr{D}X e^{-it\int_\Sigma d^2z\{Q,V\}}\mathscr{O}_1\mathscr{O}_2\cdots\mathscr{O}_n.
\end{aligned}
\tag{3.23}
$$

In the above derivation, we neglected square and higher powers of (δt) and used the relation $\int_{P_d} \mathscr{D}X e^{-it\int_\Sigma d^2z\{Q,V\}}\{Q, U\} = 0$. Hence $\langle\mathscr{O}_1\mathscr{O}_2\cdots\mathscr{O}_n\rangle_d$ does not depend on t and can be computed under the limit $t \to \infty$. In field theory, this operation is called "taking the weak coupling limit". In the weak coupling limit, saddle point approximation becomes exact (saddle point approximation means expanding the Lagrangian up to second order around the solution of the classical equation of motion and executing Gaussian integration). In the last line of (3.23), the Lagrangian $t\int_\Sigma d^2z\{Q, V\}$ is explicitly written in terms of the original fields as follows:

$$
2t\int_\Sigma d^2z\Big(g_{i\bar{j}}(\partial_z\phi^{\bar{j}}\partial_{\bar{z}}\phi^i) + ig_{i\bar{j}}\psi_z^{\bar{j}}D_{\bar{z}}\chi^i + ig_{i\bar{j}}\psi_{\bar{z}}^iD_z\chi^{\bar{j}} - R_{i\bar{j}k\bar{l}}\psi_{\bar{z}}^i\psi_z^{\bar{j}}\chi^k\chi^{\bar{l}}\Big).
\tag{3.24}
$$

For the time being, we neglect the fourth order term of fermions $-R_{i\bar{j}k\bar{l}}\psi_{\bar{z}}^i\psi_z^{\bar{j}}\chi^k\chi^{\bar{l}}$. (This operation does not affect the outline of the discussion. We recommend [2] for more rigorous discussion.) Then the classical equations of motion for the fields are given by

$$
\partial_z\phi^{\bar{i}} = 0,\ \ \partial_{\bar{z}}\phi^i = 0,\ \ D_{\bar{z}}\chi^i = 0, D_z\chi^{\bar{i}} = 0,\ \ D_z\psi_{\bar{z}}^i = 0,\ \ D_{\bar{z}}\psi_z^{\bar{i}} = 0.
\tag{3.25}
$$

Among the above equations, we focus on $\partial_{\bar{z}}\phi^i = 0, D_{\bar{z}}\chi^i = 0$ and $D_{\bar{z}}\psi_z^{\bar{i}} = 0$ because the others are obtained by taking the complex conjugate of these three equations. They have important meanings from the point of view of complex geometry and algebraic geometry. $\partial_{\bar{z}}\phi^i = 0$ is the condition for ϕ to be a holomorphic map and $D_{\bar{z}}\chi^i = 0$ is the condition for χ to be a holomorphic section of the holomorphic vector bundle $\phi^{-1}(T'M)$ on Σ, or to be an element of $H^0(\Sigma, \phi^{-1}(T'M))$. As for $D_{\bar{z}}\psi_z^{\bar{i}} = 0$, it is the condition for ψ to be an element of $H^1(\Sigma, \phi^{-1}(T'M))$. Here, we used the isomorphism of the vector bundle $\overline{T'M} \simeq (T'M)^*$ obtained from contraction by the Kähler metric and Serre duality $H^1(\Sigma, \phi^{-1}(T'M)) \simeq H^0(\Sigma, (T'\Sigma)^* \otimes \phi^{-1}((T'M)^*))$ If M is a compact Kähler manifold, the set of holomorphic maps of degree d from Riemann surface Σ to M becomes a complex space of finite dimension. (In general, it is not always a complex "manifold". Mathematically, it becomes a "stack" which has a dimensional "jump" in some locus and also admits orbifold singularities.) This space is called the "moduli space of holomorphic maps of degree d from Σ to M". Let us denote it by $\mathscr{M}_\Sigma(M,d)$ temporarily. The complex dimension of $\mathscr{M}_\Sigma(M,d)$ in a small open neighborhood of a point $\phi \in \mathscr{M}_\Sigma(M,d)$ is given by $\dim(H^0(\Sigma,\phi^{-1}(T'M))$. This is because an element of $\dim(H^0(\Sigma,\phi^{-1}(T'M))$, or a holomorphic section of $\phi^{-1}(T'M)$ can be interpreted as a holomorphic infinitesimal deformation of ϕ. On the other hand, $\dim(H^0(\Sigma,\phi^{-1}(T'M))) - \dim(H^1(\Sigma,\phi^{-1}(T'M)))$ is determined by the index theorem of the differential operator $D_{\bar{z}}$ on $\phi^{-1}(T'M)$, or the Riemann–Roch theorem as follows:

$$
\begin{aligned}
&\dim(H^0(\Sigma,\phi^{-1}(T'M))) - \dim(H^1(\Sigma,\phi^{-1}(T'M))) \\
&= \dim_{\mathbf{C}}(M)\cdot(1-g_\Sigma) + \int_M \phi^*(c_1(M)).
\end{aligned}
\tag{3.26}
$$

In (3.26), g_Σ is the genus, or the number of holes of Σ. The r.h.s. of (3.26) depends only on g_Σ, d and the topological invariant of M. Hence $\dim(H^0(\Sigma,\phi^{-1}(T'M))) - \dim(H^1(\Sigma,\phi^{-1}(T'M)))$ does not depend on ϕ and it is called the "virtual dimension" of $\mathscr{M}_\Sigma(M,d)$. Let us discuss briefly the behavior of the dimension of $\mathscr{M}_\Sigma(M,d)$. At the generic point ϕ of $\mathscr{M}_\Sigma(M,d)$, $\dim(H^1(\Sigma,\phi^{-1}(T'M)))$ vanishes and the dimension of $\mathscr{M}_\Sigma(M,d)$ in a small open neighborhood of ϕ coincides with the virtual dimension. But at a point where $\dim(H^1(\Sigma,\phi^{-1}(T'M)))$ is positive, the dimension jumps from the virtual dimension by $\dim(H^1(\Sigma,\phi^{-1}(T'M)))$. One can visualize this kind of situation by imagining a space obtained from gluing real 2-dimensional spheres at integer points of real number line. In this space, the dimension at the non-integer point of the real number line is 1, but the dimension at the integer point is given by 2. So far, we have explained how solutions of (3.25) are uniformly interpreted in terms of geometry of $\mathscr{M}_\Sigma(M,d)$. In the well-known example of classical mirror symmetry, g_Σ is 0 and M is a complex 3-dimensional Calabi–Yau manifold. Then the r.h.s. of (3.26) becomes 3 regardless of degree d because $c_1(M)=0$. This fact is deeply related to the structure of the instanton expansion of the Yukawa coupling associated with Kähler deformation of M, as will be discussed later.

Let us denote solutions of (3.25) by zeromodes, ϕ_0, χ_0 and ψ_0 and expand each field around a zeromode as follows:

$$
\begin{aligned}
&\phi^i = \phi_0^i + \phi'^i, \quad \chi^i = \chi_0^i + \chi'^i, \quad \psi_z^{\bar{i}} = \psi_{0z}^{\bar{i}} + \psi_{0z}'^{\bar{i}}, \\
&\phi^{\bar{i}} = \phi_0^{\bar{i}} + \phi'^{\bar{i}}, \quad \chi^{\bar{i}} = \chi_0^{\bar{i}} + \chi'^{\bar{i}}, \quad \psi_{\bar{z}}^{i} = \psi_{0\bar{z}}^{i} + \psi_{0\bar{z}}'^{i}.
\end{aligned}
\tag{3.27}
$$

If we expand (3.24) by using (3.27) up to second order of ϕ', χ' and ψ', we obtain

$$
2t\int_\Sigma d^2z\Big(g_{i\bar{j}}(\partial_z\phi'^{\bar{j}}\partial_{\bar{z}}\phi'^i) + ig_{i\bar{j}}\psi_z'^{\bar{j}}\partial_{\bar{z}}\chi'^i + ig_{i\bar{j}}\psi_{\bar{z}}'^{i}\partial_z\chi'^{\bar{j}}\Big).
\tag{3.28}
$$

In (3.28), $D_{\bar{z}}$ and D_z become $\partial_{\bar{z}}$ and ∂_z respectively up to second order because $\partial_{\bar{z}}\phi_0^i = \partial_z\phi_0^{\bar{i}} = 0$. Moreover, in the $t\to\infty$ limit, the radius of M becomes infinity and M can be regarded as a locally flat space. Therefore, we can replace the Kähler metric $g_{i\bar{j}}$ by $\delta_{i\bar{j}}$. In this way, the quadratic Lagrangian is given as follows:

$$L_{quad.} := 2t \int_{\Sigma} d^2z \left(\sum_{i=1}^{\dim_{\mathbb{C}}(M)} (\partial_z \phi'^{\bar{i}} \partial_{\bar{z}} \phi'^{i}) + i\psi'^{\bar{i}}_z \partial_{\bar{z}} \chi'^{i} + i\psi'^{i}_{\bar{z}} \partial_z \chi'^{\bar{i}} \right). \quad (3.29)$$

The result of Gaussian integration of ϕ', χ' and ψ' is given by

$$\frac{(\det'(\partial_z))^{\dim_{\mathbb{C}}(M)} (\det'(\partial_{\bar{z}}))^{\dim_{\mathbb{C}}(M)}}{(\det'(\partial_z \partial_{\bar{z}}))^{\dim_{\mathbb{C}}(M)}} = \frac{\det'(\partial_z \partial_{\bar{z}}))^{\dim_{\mathbb{C}}(M)}}{\det'(\partial_z \partial_{\bar{z}}))^{\dim_{\mathbb{C}}(M)}} = 1, \quad (3.30)$$

where $\det'(A)$ is the product of non-zero eigenvalues of A. In sum, contributions of oscillation modes of bosons and fermions cancel each other. This is one of major characteristics of topological field theory.

Let us summarize the discussions of this section. In evaluating the degree d correlation function,

$$\langle \mathcal{O}_1 \mathcal{O}_2 \cdots \mathcal{O}_n \rangle_d = \int_{P_d} \mathcal{D}X e^{-it\int_\Sigma d^2z\{Q,V\}} \mathcal{O}_1 \mathcal{O}_2 \cdots \mathcal{O}_n, \quad (3.31)$$

we decomposed the field measure $\mathcal{D}X = \mathcal{D}\phi \mathcal{D}\chi \mathcal{D}\psi$ into a zeromode part $\mathcal{D}\phi_0 \mathcal{D}\chi_0 \mathcal{D}\psi_0$ and an oscillation mode part $\mathcal{D}\phi' \mathcal{D}\chi' \mathcal{D}\psi'$ and applied saddle point approximation under the weak coupling limit. Accordingly, the Lagrangian $it \int_\Sigma d^2z\{Q, V\}$ is decomposed into the classical part $L_{cl.}$ and the quadratic part $L_{quad.}$. In this way, (3.31) is rewritten as follows:

$$\langle \mathcal{O}_1 \mathcal{O}_2 \cdots \mathcal{O}_n \rangle_d = \int_{\mathcal{M}_\Sigma(M,d)} \mathcal{D}\phi_0 \mathcal{D}\chi_0 \mathcal{D}\psi_0 e^{-L_{cl.}} \left(\int \mathcal{D}\phi' \mathcal{D}\chi' \mathcal{D}\psi' e^{-L_{quad.}} \right) \mathcal{O}_1 \mathcal{O}_2 \cdots \mathcal{O}_n.$$

Since the result of Gaussian integration turns out to be 1, it becomes

$$\langle \mathcal{O}_1 \mathcal{O}_2 \cdots \mathcal{O}_n \rangle_d = \int_{\mathcal{M}_\Sigma(M,d)} \mathcal{D}\phi_0 \mathcal{D}\chi_0 \mathcal{D}\psi_0 e^{-L_{cl.}} \mathcal{O}_1 \mathcal{O}_2 \cdots \mathcal{O}_n. \quad (3.32)$$

As for $L_{cl.}$, it is created from the term:

$$-2t \int_\Sigma d^2z R_{\bar{i}jk\bar{l}} \psi^i_{\bar{z}} \psi^{\bar{j}}_z \chi^k \chi^{\bar{l}}, \quad (3.33)$$

which has been neglected so far, and it vanishes when there are no ψ zero modes.

3.2.2 *Observable Satisfying* $\{Q, \mathcal{O}\} = 0$ *and Topological Selection Rule*

In this subsection, we explicitly construct the observable $\mathcal{O}$ that satisfies the condition $\{Q, \mathcal{O}\} = 0$ in the previous section. As one can imagine from analogy between $Q^2 =$

0 and $d^2 = 0$, it is constructed from an element of the cohomology ring of M. Let us introduce a differential form W of bi-degree (p, q) on the Kähler manifold M as follows:

$$W = W_{i_1 \cdots i_p \bar{j}_1 \cdots \bar{j}_q}(x) dx^{i_1} \wedge \cdots \wedge dx^{i_p} \wedge dx^{\bar{j}_1} \wedge \cdots \wedge dx^{\bar{j}_q}. \tag{3.34}$$

We then consider an observable $\mathcal{O}_W(z)$ associated with W:

$$\mathcal{O}_W(z) = W_{i_1 \cdots i_p \bar{j}_1 \cdots \bar{j}_q}(\phi(z)) \chi^{i_1}(z) \cdots \chi^{i_p}(z) \chi^{\bar{j}_1}(z) \cdots \chi^{\bar{j}_q}(z). \tag{3.35}$$

From the BRST transformation given in (3.8),

$$\{Q, \phi^i\} = \chi^i, \ \{Q, \phi^{\bar{i}}\} = \chi^{\bar{i}}, \ \{Q, \chi^i\} = 0, \ \{Q, \chi^{\bar{i}}\} = 0, \tag{3.36}$$

we obtain the following relation:

$$\{Q, \mathcal{O}_W(z)\} = \mathcal{O}_{dW}(z). \tag{3.37}$$

Therefore, $\{Q, \mathcal{O}_W(z)\} = 0$ is automatically satisfied if W is a closed form. Moreover, if W is an exact form, or equivalently, there exists a differential form V that satisfies $W = dV$, the relation $\mathcal{O}_W(z) = \{Q, \mathcal{O}_V(z)\}$ holds. As we have explained in the previous section, the degree d correlation function $\langle \mathcal{O}_1 \mathcal{O}_2 \cdots \mathcal{O}_n \{Q, \mathcal{O}_V(z)\} \rangle_d$ vanishes if $\{Q, \mathcal{O}_i\} = 0, \ (i = 1, 2, \ldots, n)$ hold. Hence the operator $\{Q, \mathcal{O}_V(z)\}$ has no meaning in this theory. In sum, the physical observable is given by $\mathcal{O}_W(z)$, where W is an element of the cohomology ring of M. From now on, we consider the degree d correlation function given by

$$\langle \mathcal{O}_{W_1}(z_1) \mathcal{O}_{W_2}(z_2) \cdots \mathcal{O}_{W_n}(z_n) \rangle_d \quad (W_I \in H^{p_I, q_I}(M, \mathbf{C})). \tag{3.38}$$

Let us write down the path-integral representation of the above correlation function obtained in the previous section:

$$\langle \mathcal{O}_{W_1}(z_1) \mathcal{O}_{W_2}(z_2) \cdots \mathcal{O}_{W_n}(z_n) \rangle_d = \int_{\mathcal{M}_{\Sigma}(M,d)} \mathcal{D}\phi_0 \mathcal{D}\chi_0 \mathcal{D}\psi_0 e^{-L_{cl.}} \mathcal{O}_{W_1}(z_1) \mathcal{O}_{W_2}(z_2) \cdots \mathcal{O}_{W_n}(z_n). \tag{3.39}$$

Since all the oscillation modes are integrated out in the above representation, $\mathcal{O}_{W_I}(z)$ takes the form

$$\mathcal{O}_{W_I}(z) = W_{I, i_1 \cdots i_{p_I} \bar{j}_1 \cdots \bar{j}_{q_I}}(\phi_0(z)) \chi_0^{i_1}(z) \cdots \chi_0^{i_{p_I}}(z) \chi_0^{\bar{j}_1}(z) \cdots \chi_0^{\bar{j}_{q_I}}(z). \tag{3.40}$$

In other words, all the fields are replaced by their zero modes. At this stage, let us focus on the fermion measure $\mathcal{D}\chi_0 \mathcal{D}\psi_0$. As we have seen in the previous section, the number of solutions of $D_{\bar{z}}\chi^i = 0$ is given by $a_d = \dim(H^0(\Sigma, \phi^{-1}(T'M)))$. Since $D_z \chi^{\bar{i}} = 0$ is complex conjugate of $D_{\bar{z}}\chi^i = 0$, the number of its solutions is also

a_d. We denote these zero modes by $\chi_{0,a}$ and $\overline{\chi}_{0,a}$ ($a = 1, 2, \ldots, a_d$). We also have $b_d = \dim(H^1(\Sigma, \phi^{-1}(T'M)))$ solutions of $D_{\bar{z}}\psi^{\bar{i}}_z = 0$ and $D_z\psi^{i}_{\bar{z}} = 0$ respectively, and we denote these zero modes by $\psi_{0,b}$ and $\overline{\psi}_{0,b}$ ($b = 1, 2, \ldots, b_d$). With these notations, $\mathscr{D}\chi_0\mathscr{D}\psi_0$ is explicitly written as the corresponding measure of Grassmann numbers:

$$d\chi_{0,1}d\overline{\chi}_{0,1}\cdots d\chi_{0,a_d}d\overline{\chi}_{0,a_d}d\psi_{0,1}d\overline{\psi}_{0,1}\cdots d\psi_{0,b_d}d\overline{\psi}_{0,b_d}. \tag{3.41}$$

To obtain a non-vanishing correlation function, we have to supply the same number of zero modes from $e^{-L_{cl.}}\mathscr{O}_{W_1}(z_1)\mathscr{O}_{W_2}(z_2)\cdots\mathscr{O}_{W_n}(z_n)$. Equation (3.40) tells us that we can obtain p_I χ zero modes, q_I $\overline{\chi}$ zero modes and no ψ and $\overline{\psi}$ zero modes from $\mathscr{O}_{W_I}(z)$. On the other hand, since $-L_{cl.}$ is created from the term $R_{\bar{i}j k\bar{l}}\psi^{\bar{i}}_{0\bar{z}}\psi^{j}_{0z}\chi^{k}_{0}\chi^{\bar{l}}_{0}$, the same number of χ, $\overline{\chi}$, ψ and $\overline{\psi}$ zero modes are supplied by $e^{-L_{cl.}}$. With these things considered, the conditions

$$a_d - b_d \geq 0, \tag{3.42}$$

and

$$\sum_{I=1}^{n} p_I = a_d - b_d, \quad \sum_{I=1}^{n} q_I = a_d - b_d, \tag{3.43}$$

must hold to obtain a non-vanishing degree d correlation function. The condition given by (3.43) is called the "topological selection rule" of a degree d correlation function.

3.2.3 *Geometrical Interpretation of Degree d Correlation Function*

Let us explain the geometrical meaning of the correlation function $\langle\mathscr{O}_{W_1}(z_1)\cdots\mathscr{O}_{W_n}(z_n)\rangle_d$ that satisfies the topological selection rule. We introduce here the notion of Poincaré duality. Let M be a complex n-dimensional compact Kähler manifold. The Poincaré dual $PD(W)$ of a cohomology element $W \in H^{p,q}(M, \mathbf{C})$ is a submanifold of M that satisfies the following condition:

$$\int_M W \wedge \alpha = \int_{PD(W)} i^*(\alpha), \tag{3.44}$$

where $i : PD(W) \hookrightarrow M$ is an inclusion map and α is an arbitrary element of $H^{*,*}(M, \mathbf{C})$. It has real dimension $2n - (p + q)$ (or real codimension $p + q$) and is determined up to homology equivalence. Let us fix a representative element of $PD(W)$ for each $W \in H^{*,*}(M, \mathbf{C})$. By the localization principle of mathematical theory of differential forms, we can choose a differential form which has delta-function support on $PD(W)$, as the representative differential form of W.

Let us consider the simple case where $b_d = 0$, i.e., there exist no ψ zero modes. In this case, $L_{cl} = 0$ and the degree d correlation is written as follows:

$$\langle \mathcal{O}_{W_1}(z_1)\mathcal{O}_{W_2}(z_2)\cdots\mathcal{O}_{W_n}(z_n)\rangle_d = \int_{\mathcal{M}_\Sigma(M,d)} \mathcal{D}\phi_0 d\chi_{0,1} d\overline{\chi}_{0,1}\cdots d\chi_{0,a_d} d\overline{\chi}_{0,a_d} \mathcal{O}_{W_1}(z_1)\mathcal{O}_{W_2}(z_2)\cdots\mathcal{O}_{W_n}(z_n). \tag{3.45}$$

In the above representation, we take each W_I in $\mathcal{O}_{W_I}(z_I)$ as the differential form having delta function support on $PD(W_I)$. As one can see from the explicit form of $\mathcal{O}_{W_I}(z_I)$,

$$\mathcal{O}_{W_I}(z_I) = W_{I,i_1\cdots i_{p_I}\bar{j}_1\cdots\bar{j}_{q_I}}(\phi_0(z_I))\chi_0^{i_1}(z)\cdots\chi_0^{i_{p_I}}(z)\chi_0^{\bar{j}_1}(z)\cdots\chi_0^{\bar{j}_{q_I}}(z), \tag{3.46}$$

integrating out ϕ_0 on the moduli space $\mathcal{M}_\Sigma(M, d)$ corresponds to picking up holomorphic maps ϕ_0 that satisfy the following condition:

$$\phi_0(z_I) \in PD(W_I), \quad (I = 1, 2, \ldots, n). \tag{3.47}$$

For each I, the condition (3.47) forces the fixed point z_I on Σ to be mapped to the submanifold $PD(W_I)$ of codimension $p_I + q_I$. Hence it imposes $p_I + q_I$ constraints on degrees of freedom of movement of ϕ_0. With all the I's considered, (3.47) imposes $\sum_{I=1}(p_I + q_I)$ constraints on ϕ_0. The topological selection rule tells us that this sum equals $2a_d$, which is nothing but the real dimension of $\mathcal{M}_\Sigma(M, d)$ (we assume here $b_d = 0$). This means that the set of holomorphic maps satisfying (3.47) becomes a discrete set. After all, integrating out ϕ_0 on $\mathcal{M}_\Sigma(M, d)$ results in counting the number of ϕ_0's that satisfy (3.47). As for fermions, all the χ_0's appearing in $\mathcal{O}_{W_I}(z_I)$ are adequately soaked up by χ_0 measures by topological selection rule and the result of fermion integration turns out to be 1. In sum, we have come to the following conclusion.

If there exist no ψ zero modes, the degree d correlation function $\langle \mathcal{O}_{W_1}(z_1)\cdots \mathcal{O}_{W_n}(z_n)\rangle_d$ equals the number of holomorphic maps $\phi_0 : \Sigma \to M$ that satisfy the condition $\phi_0(z_I) \in PD(W_I), \quad (I = 1, 2, \ldots, n)$.

Next, we consider the case when the number of ψ zero modes b_d is non-zero. Also in this case, $\mathcal{O}_{W_1}(z_1)\mathcal{O}_{W_2}(z_2)\cdots\mathcal{O}_{W_n}(z_n)$ imposes the condition (3.47) on ϕ_0, but the number of constraints is $\sum_{I=1}(p_I + q_I) = 2(a_d - b_d)$, as can be seen from the topological selection rule. Hence there remain $2b_d$ degrees of freedom of ϕ_0 since $\dim_{\mathbf{R}}(\mathcal{M}_\Sigma(M, d))$ is $2a_d$. Let us denote by $\mathcal{M}^R_\Sigma(M, d)$ the space created by the remaining $2b_d$ degrees of freedom. At this stage, the factor $e^{-L_{cl.}}$ plays an important role. Rigorous treatment of the path integral tells us that it becomes the top Chern class $c_{b_d}(\mathcal{H}^1)$ of the rank b_d vector bundle $\mathcal{H}^1$ on $\mathcal{M}^R_\Sigma(M, d)$. Since the fiber of $\mathcal{H}^1$ on a point $\phi_0 \in \mathcal{M}^R_\Sigma(M, d)$ is given by $H^1(\Sigma, \phi_0^{-1}(T'M))$, the rank of $\mathcal{H}^1$ equals $b_d = \dim(H^1(\Sigma, \phi_0^{-1}(T'M)))$. $c_{b_d}(\mathcal{H}^1)$ includes, b_d χ_0's, $\overline{\chi}_0$'s, ψ_0's and $\overline{\psi}_0$'s respectively, and these are adequately soaked up by Grassmann measures that survived integration of fermion zero modes of $\mathcal{O}_{W_1}(z_1)\mathcal{O}_{W_2}(z_2)\cdots\mathcal{O}_{W_n}(z_n)$. In sum, we obtain the following representation of the degree d correlation function:

$$\langle \mathcal{O}_{W_1}(z_1)\mathcal{O}_{W_2}(z_2)\cdots\mathcal{O}_{W_n}(z_n)\rangle_d = \int_{\mathcal{M}^R_\Sigma(M,d)} c_{b_d}(\mathcal{H}^1). \tag{3.48}$$

When there exist no ψ zero modes, i.e., $b_d = 0$, $\mathcal{M}^R_\Sigma(M, d)$ is a discrete set. In this case, if we think that $c_{b_d}(\mathcal{H}^1)$ is given by 1 and that integration corresponds to counting the number of elements of the discrete set, (3.48) can be regarded as the extension of geometrical interpretation of the degree d correlation function in the $b_d = 0$ case. We end here discussions of geometrical interpretation of degree d correlation functions in this subsection. In the next subsection, we pick up a degree d correlation function of a degree 5 hypersurface in CP^4, the complex 3-dimensional Calabi–Yau manifold used in the celebrated paper by Candelas et al., as a specific example and explain how the formula (3.48) is given concrete shape.

3.2.4 Degree d Correlation Function of Degree 5 Hypersurface in CP^4

In this subsection, we take a degree 5 hypersurface in CP^4, which we denote by M^5_4, as an example of a target Kähler manifold M and CP^1 as an example of a Riemann surface Σ. In this setting, we take a closer look at the degree d correlation function introduced in the previous subsection. As we have discussed in Chap. 2, CP^4 is the complex 4-dimensional manifold whose points are described by ratio of five complex homogeneous coordinates X_i $(i = 1, 2\cdots, 5)$:

$$CP^4 := \{(X_1 : X_2 : X_3 : X_4 : X_5) \mid X_i \in \mathbf{C}\}. \tag{3.49}$$

M^5_4 is the complex 3-dimensional submanifold of CP^4 which is defined as the zero locus of a degree 5 homogeneous polynomial in X_i's. For simplicity, we take the homogeneous polynomial in the following form:

$$M^5_4 := \{(X_1 : X_2 : X_3 : X_4 : X_5) \in CP^4 \mid \sum_{j=1}^{5}(X_j)^5 = 0\}. \tag{3.50}$$

Let s and t be two homogeneous coordinates that describe points of CP^1:

$$CP^1 := \{\, (s : t) \mid s, t \in \mathbf{C}\}, \tag{3.51}$$

CP^1 can be topologically regarded as a real 2-dimensional sphere S^2 since it is obtained by adding a point of infinity $(0 : 1)$ to the complex plane $\{(1 : z) \mid z \in \mathbf{C}\}$. As we have seen in the previous subsection, evaluation of the degree d correlation function is directly connected to the structure of the moduli space of degree d holomorphic maps from CP^1 to M_4^5, which we denote by $\mathcal{M}_{CP^1}(M_4^5, d)$. It is not straightforward to tackle the geometrical structure of $\mathcal{M}_{CP^1}(M_4^5, d)$. Therefore, we begin by analyzing the geometrical structure of $\mathcal{M}_{CP^1}(CP^4, d)$, the moduli space of degree d holomorphic maps from CP^1 to CP^4 because a holomorphic map from CP^1 to M_4^5 is given by a holomorphic map from CP^1 to CP^4 whose image is contained in M_4^5. As is well known in algebraic geometry, a degree d holomorphic map ϕ_0 from CP^1 to CP^4 is always represented by 5-tuple of degree d homogeneous polynomials in s and t as follows:

$$\phi_0(s:t) = (\sum_{j=0}^{d} a_j^1 s^j t^{d-j} : \sum_{j=0}^{d} a_j^2 s^j t^{d-j} : \sum_{j=0}^{d} a_j^3 s^j t^{d-j} : \sum_{j=0}^{d} a_j^4 s^j t^{d-j} : \sum_{j=0}^{d} a_j^5 s^j t^{d-j}). \quad (3.52)$$

We explain here the geometrical background to this representation. The degree d of a map ϕ_0 was originally defined by the relation

$$\int_{CP^1} \phi_0^*(K) = 2\pi i d, \quad (3.53)$$

where $K = g_{i\bar{j}} dz^i \wedge d\bar{z}^j$ is the Kähler form of CP^4. If we set $h := \frac{1}{2\pi i} K$, the above relation reduces to

$$\int_{CP^1} \phi_0^*(h) = d. \quad (3.54)$$

As we have seen in the discussion in Sect. 2.4, this h is the first Chern class $c_1(H)$ of the hyperplane bundle H on CP^4. By combining this fact with the Gauss–Bonnet theorem of Chern classes, we can translate (3.54) into the following statement.

"Let $s : CP^4 \to H$ be a global holomorphic section of H. Then the number of zero points (counted with multiplicity) of $\phi_0^*(s) : CP^1 \to H$, the pull-back of s by ϕ_0, equals d."

The global section of H is given by a degree 1 homogeneous polynomial in X_i's. So let us take X_i $(i = 1, \ldots, 5)$ as global sections. Since ϕ_0 is a holomorphic map from CP^1 to CP^4, $\phi_0^*(X_i)$ must be a homogeneous polynomial in s and t. Then the above condition tells us that its homogeneous degree is given by d (cf. the fundamental theorem of algebra):

$$\phi_0^*(X_i) = \sum_{j=0}^{d} a_j^i s^j t^{d-j}. \quad (3.55)$$

The fact that X_i's are homogeneous coordinates of CP^4 leads us to

$$\phi_0(s:t) = (\phi_0^*(X_1) : \phi_0^*(X_2) : \phi_0^*(X_3) : \phi_0^*(X_4) : \phi_0^*(X_5)). \quad (3.56)$$

In this way, we obtain the representation (3.52).

Now, we heuristically assume that the representation in the r.h.s. of (3.52) always defines a degree d holomorphic map from CP^1 to CP^4. Then the holomorphic is parametrized by its coefficients (a^i_j). On the other hand, multiplying each coefficient a^i_j by the same complex number $\lambda \in \mathbf{C}^{\times}$ does not change the map because of projective equivalence. Hence it seems that $\mathcal{M}_{CP^1}(CP^4, d)$ is given by

$$CP^{5(d+1)-1} := \{(a^1_0 : a^1_1 : \cdots : a^1_d : a^2_0 : a^2_1 : \cdots : a^5_d) \mid a^i_j \in \mathbf{C}\}. \quad (3.57)$$

This space has a complex dimension that coincides with the virtual dimension of $\mathcal{M}_{CP^1}(CP^4, d)$ obtained from the Riemann–Roch theorem:

$$\dim_{\mathbf{C}}(CP^4) + \int_{CP^1} \phi_0^* c_1(CP^4) = 4 + \int_{CP^1} \phi_0^*(5h) = 5d + 4. \quad (3.58)$$

But from a rigorous point of view, it should be regarded as the first approximation of the moduli space. This is because the representation in the r.h.s. of (3.52) must satisfy the following condition in order to define a holomorphic map from CP^1 to CP^4:

$\sum_{j=0}^{d} a^i_j s^j t^{d-j}$ $(i = 1, \ldots, 5)$ do not have a homogeneous polynomial

of positive degree $\sum_{j=0}^{f} b_j s^j t^{f-j}$ as a common divisor. (3.59)

Let us assume that the r.h.s. of (3.52) does not satisfy (3.59). Then there exists a degree f homogeneous polynomial $\sum_{j=0}^{f} b_j s^j t^{f-j}$ and each $\sum_{j=0}^{d} a^i_j s^j t^{d-j}$ is factorized as follows:

$$\sum_{j=0}^{d} a^i_j s^j t^{d-j} = (\sum_{j=0}^{f} b_j s^j t^{f-j}) \cdot (\sum_{j=0}^{d-f} c^i_j s^j t^{d-f-j}), \ (i = 1, \ldots, 5), (1 \leq f \leq d). \quad (3.60)$$

In (3.60), we require $\sum_{j=0}^{d-f} c^i_j s^j t^{d-f-j}$ $(i = 1, \ldots, 5)$ to have no common divisors of positive degree. If we factorize $(\sum_{j=0}^{f} b_j s^j t^{f-j})$ into $\prod_{j=1}^{f}(\beta^j_2 s - \beta^j_1 t)$, we can observe that the r.h.s. of (3.52) becomes $(0:0:0:0:0)$ at $(s:t) = (\beta^j_1 : \beta^j_2)$, $(j = 1, \ldots, f)$. Therefore, the r.h.s. of (3.52) does not define a map from CP^1 to CP^4 at a point in $CP^{5(d+1)-1}$ where the condition (3.60) holds. In this way, we obtain the following description of $\mathcal{M}_{CP^1}(CP^4, d)$:

$$\mathcal{M}_{CP^1}(CP^4, d) = CP^{5(d+1)-1} - \{\text{the set of points in } CP^{5(d+1)-1} \text{ that satisfy(3.60)}\},$$

and inclusion map $i_0 : \mathcal{M}_{CP^1}(CP^4, d) \hookrightarrow CP^{5(d+1)-1}$. Next, let us focus on the points that belong to $CP^{5(d+1)-1} - \mathcal{M}_{CP^1}(CP^4, d)$. The homogeneous polynomials $\sum_{j=0}^{d} a^i_j s^j t^{d-j}$ $(i = 1, \ldots, 5)$ that are obtained from these points are always factorized in the form given in (3.60). Since we have required $\sum_{j=0}^{d-f} c_j^i s^j t^{d-f-j}$ $(i = 1, \ldots, 5)$ to have no common divisor of positive degree, the representation

$$(\sum_{j=0}^{d-f} c_j^1 s^j t^{d-f-j} : \sum_{j=0}^{d-f} c_j^2 s^j t^{d-f-j} : \cdots : \sum_{j=0}^{d-f} c_j^5 s^j t^{d-f-j}), \tag{3.61}$$

defines a degree $d-j$ map from CP^1 to CP^4. In other words, $(c_0^1 : c_1^1 : \cdots : c_{d-f}^5) \in CP^{5(d-f+1)-1}$ becomes a point in $\mathcal{M}_{CP^1}(CP^4, d-f)$. Conversely, if we introduce $CP^f := \{(b_0 : b_1 : \cdots : b_f) \mid b_j \in \mathbf{C}\}$ to parametrize the degree j homogeneous polynomial $(\sum_{j=0}^{f} b_j s^j t^{f-j})$ (identified up to multiplication by $\lambda \in \mathbf{C}^\times$), we can construct a map,

$$i_f : CP^f \times \mathcal{M}_{CP^1}(CP^4, d-f) \to CP^{5(d+1)-1}, \quad (f = 1, \ldots, d), \tag{3.62}$$

by picking up the coefficients of degree d homogeneous polynomials obtained from expanding the r. h. s. of (3.60). This map provides embedding $i_f : CP^f \times \mathcal{M}_{CP^1}(CP^4, d-f) \hookrightarrow CP^{5(d+1)-1}$. Moreover, factorization given in (3.60) is unique up to multiplication by a non-zero constant. Hence we obtain the following decomposition:

$$CP^{5(d+1)-1} - \mathcal{M}_{CP^1}(CP^4, d) = \coprod_{f=1}^{d} (CP^f \times \mathcal{M}_{CP^1}(CP^4, d-f)). \tag{3.63}$$

Since $\mathcal{M}_{CP^1}(CP^4, d-f)$ is contained in $CP^{5(d-f+1)-1}$, the complex dimension of the spaces appearing in the r.h.s. is less than $5d+4$, which is the complex dimension of $CP^{5(d+1)-1}$. This allows us to conclude that the complex dimension of $\mathcal{M}_{CP^1}(CP^4, d)$ equals $5d+4$. It also equals the virtual dimension of $\mathcal{M}_{CP^1}(CP^4, d)$ and no ψ-zero modes appear in this case.

Now, let us turn our attention to $\mathcal{M}_{CP^1}(M_4^5, d)$, the moduli space of degree d holomorphic maps from CP^1 to M_4^5. Since M_4^5 is a submanifold in CP^4, a holomorphic map ϕ_0 from CP^1 to M_4^5 is given by the holomorphic map from CP^1 to CP^4 that satisfies $\phi_0(CP^1) \subset M_4^5$. This condition is equivalent to the condition that $\phi_0(s:t)$ satisfies the defining equation $(X_1)^5 + (X_2)^5 + \cdots + (X_5)^5 = 0$ of M_4^5 for an arbitrary $(s:t) \in CP^1$. Therefore, the equation

$$(\sum_{j=0}^{d} a_j^1 s^j t^{d-j})^5 + (\sum_{j=0}^{d} a_j^2 s^j t^{d-j})^5 + \cdots + (\sum_{j=0}^{d} a_j^5 s^j t^{d-j})^5 = 0 \tag{3.64}$$

must hold for an arbitrary $(s:t)$. Here we use the representation (3.52) of ϕ_0. By expanding the above equation in s and t, it reduces to

$$\sum_{m=0}^{5d} f_m^5(a_j^i)s^m t^{5d-m} = 0, \tag{3.65}$$

where $f_m^5(a_j^i)$ $(m = 0, 1, \ldots, 5d)$ are degree 5 homogeneous polynomials in a_j^i. Since (3.65) must hold for an arbitrary $(s : t)$, the condition is translated into the following condition of (a_j^i):

$$f_m^5(a_j^i) = 0, \quad (m = 0, 1, \ldots, 5d), \tag{3.66}$$

Optimistically, we assume that these $5d + 1$ equations are independent. Then they impose $5d + 1$ constraints on variation of motion of ϕ_0. Since the complex dimension of $\mathcal{M}_{CP^1}(CP^4, d)$ is $5d + 4$, the expected complex dimension of $\mathcal{M}_{CP^1}(M_4^5, d)$ is given by $5d + 4 - (5d + 1) = 3$. Indeed, this dimension coincides with the virtual dimension of $\mathcal{M}_{CP^1}(M_4^5, d)$, evaluated from the Rieman–Roch theorem,

$$\dim_{\mathbf{C}}(M_4^5) + \int_{CP^1} \phi_0^*(c_1(M_4^5)) = 3 + 0 = 3. \tag{3.67}$$

Here, we used the fact that $c_1(M_4^5)$ vanishes. At this stage, we introduce the Clemens conjecture, which is a conjecture in algebraic geometry.

Conjecture 3.2.1 *We call a $CP^1 \subset M_4^5$ embedded by degree $d(> 0)$ holomorphic map $\phi_0 : CP^1 \to M_4^5$ a degree d rational curve in M_4^5. Then a degree d rational curve in M_4^5 does not have continuous degrees of freedom of deformation. Moreover, the set of degree d rational curves in M_4^5 becomes a finite discrete set.*

At first sight, this conjecture seems to contradict the above discussion, because it looks like asserting that the complex dimension of the moduli space should be 0. But we have to take account of the difference between deformation of a holomorphic map ϕ_0 and deformation of the "image" of ϕ_0. The moduli space $\mathcal{M}_{CP^1}(M_4^5, d)$ focuses on the former and the Clemens conjecture focuses on the latter. Let ϕ_0 be a degree d holomorphic map and also be an embedding (in particular one-to-one). According to the conjecture, $\phi_0(CP^1) := C_d \subset M_4^5$ is fixed and it can be regarded as a fixed CP^1. Therefore, the continuous deformation space of the map ϕ_0 is given by the moduli space of one-to-one holomorphic maps from CP^1 to $CP^1 = C_d$. Since the condition one-to-one is equivalent to degree 1 in this case, we can denote it by $\mathcal{M}_{CP^1}(CP^1, 1)$. In the same manner as (3.52), the degree 1 holomorphic map m_1 from CP^1 to CP^1 is always written in the form

$$m_1(s : t) = (c_0^1 t + c_1^1 s : c_0^2 t + c_1^2 s), \quad (c_0^1 c_1^2 - c_1^1 c_0^2 \neq 0), \tag{3.68}$$

where the condition in parentheses comes from the condition that m_1 is one-to-one. The set of maps given by (3.68) becomes a group whose product is defined by the composition of the maps and it is denoted by $PGL(2, \mathbf{C})$. Hence the moduli space $\mathcal{M}_{CP^1}(CP^1, 1)$ turns out to be $PGL(2, \mathbf{C})$. Since an element of $PGL(2, \mathbf{C})$

is represented by $(c_0^1 : c_1^1 : c_0^2 : c_1^2)$, $(c_0^1c_1^2 - c_1^1c_0^2 \neq 0)$, the complex dimension of $PGL(2, \mathbf{C})$ is given by $4 - 1 = 3$. In this way, if we believe in the conjecture, we can regard the 3 in the previous discussion as the complex dimension 3 of $PGL(2, \mathbf{C})$. Let us assume temporarily that $\mathcal{M}_{CP^1}(M_4^5, d)$ consists of degree d one-to-one maps from CP^1 to rigid degree d rational curves in M_4^5. Let n_d be the number of rigid degree d rational curves in M_4^5. Then $\mathcal{M}_{CP^1}(M_4^5, d)$ turns out to be n_d copies of $PGL(2, \mathbf{C})$ and its dimension is exactly 3. This argument seems to give us a clear geometrical picture of M_4^5, but unfortunately, our assumption is not correct. There exists a holomorphic map $\phi_0 \in \mathcal{M}_{CP^1}(M_4^5, d)$ which is not one-to-one. Let f be a divisor of d which does not equal 1. Then a map $\tilde{\phi}_0 \circ m_f$, which is given by the composition of a degree d holomorphic map m_f from CP^1 to CP^1 and a one-to-one holomorphic map from CP^1 to a degree $\frac{d}{f}$ rational curve, also becomes an element of $\mathcal{M}_{CP^1}(M_4^5, d)$. We can easily confirm this fact by using the explicit form of m_f,

$$m_f(s : t) = (\sum_{j=0}^{f} c_j^1 s^j t^{f-j} : \sum_{j=0}^{f} c_j^2 s^j t^{f-j}). \tag{3.69}$$

In this case, a connected component of the moduli space arising from a degree $\frac{d}{f}$ rational curve is isomorphic to $\mathcal{M}_{CP^1}(CP^1, f)$. As can be seen from (3.69), the complex dimension of $\mathcal{M}_{CP^1}(CP^1, f)$ is given by $2f + 1$ and it exceeds the virtual dimension by $2f + 1 - 3 = 2f - 2$. Hence $2f - 2$ ψ zero modes appear on this component. Fortunately, the types of holomorphic maps in $\mathcal{M}_{CP^1}(M_4^5, d)$ are exhausted by the discussions so far and the connected components of $\mathcal{M}_{CP^1}(M_4^5, d)$ are classified as follows (in the case that the Clemens conjecture holds true):

(a) n_d copies of $PGL(2, \mathbf{C})$.
(b) $n_{\frac{d}{f}}$ copies of $\mathcal{M}_{CP^1}(CP^1, f)$, $(f|d$ and $f > 1)$.

We have clarified the geometrical picture of the moduli space $\mathcal{M}_{CP^1}(M_4^5, d)$, and we turn to discussion on the degree d correlation function of M_4^5. We pick up the specific correlation function $\langle \mathcal{O}_h(z_1)\mathcal{O}_h(z_2)\mathcal{O}_h(z_3)\rangle_d$, where h is the (1, 1) form, which is given as the first Chern class of the hyperplane bundle on CP^4 (to be more precise, it should be written as $i^*(h)$ by using the inclusion map $i : M_4^5 \hookrightarrow CP^4$, but we simply denote it by h for brevity). Since h coincides with the Kähler form up to constant multiplication, it can be identified with the Kähler class as an element of the cohomology ring. Since this correlation function satisfies the topological selection rule (3.43), it can have a non-zero value. In order to evaluate it, we only have to sum up contributions from connected components labeled by degree d (or degree $\frac{d}{f}$) rational curves in M_4^5.

First, we evaluate the contribution from a connected component of the type (a) above. In this case, we have no ψ zero modes and we only have to consider the condition (3.47). In other words, it is enough for us to count holomorphic maps ϕ_0 that satisfy,

$$\phi_0(z_i) \in PD(h_i) \quad (i = 1, 2, 3), \tag{3.70}$$

in the connected component. Here, we labeled h by subscript i to distinguish h's that belong to different observables. Any holomolophic map ϕ_0 in the connected component is represented as $\phi_{0f} \circ m_1$, the composition of a fixed holomorphic map ϕ_{0f} and a holomorphic map $m_1 : CP^1 \to CP^1$ that belongs to $PGL(2, \mathbf{C})$. Hence substantial degrees of freedom of the component are described by $m_1 \in PSL(2, C)$. Let us take a closer look at the condition (3.70). As mentioned before, $PD(h_i)$ is realized as some hyperplane H_i in CP^4 ($M_4^5 \cap H_i$, to be precise). On the other hand, $\phi_0(CP^1) = \phi_{0f}(CP^1)$ is a degree d rational curve in CP^4 (also contained in M_4^5), and it intersects with each H_i in d points. We denote these intersection points by p_j^i $(i = 1, 2, 3, j = 1, 2, \ldots, d)$. If we take the hyperplanes H_i adequately, we can make these $3d$ points distinct from each other. Since ϕ^{0f} is one-to-one, inverse images $\phi_{0f}^{-1}(p_j^i)$ $(i = 1, 2, 3, j = 1, 2, \ldots, d)$ become $3d$ distinct points in CP^1. With this set-up, the condition (3.70) is rewritten as follows.

$$m_1(z_i) = \phi_{0f}^{-1}(p_{j_i}^i) \; (i = 1, 2, 3), \; j_i \in \{1, 2, \ldots, d\}. \tag{3.71}$$

Let us introduce here a well-known characteristic of $PGL(2, \mathbf{C})$ (its proof is an easy exercise).

- For a 2-tuple of 3 distinct points in CP^1, $\{z_1, z_2, z_3\}$ and $\{w_1, w_2, w_3\}$, there exists an unique element $m_1 \in PSL(2, C)$ that satisfies $m_1(z_i) = w_i$ $(i = 1, 2, 3)$.

By using this fact, we can find d^3 holomorphic maps that satisfy (3.71) for d^3 choices of (j_1, j_2, j_3). Hence the contribution from a connected component of type (a) to $\langle \mathcal{O}_h(z_1)\mathcal{O}_h(z_2)\mathcal{O}_h(z_3)\rangle_d$ is given by d^3.

Next, let us evaluate the contribution from a connected component of type (b). In this case, we have $2f - 2$ ψ zero modes, which makes evaluation more complicated. The image of ϕ_0 is now a degree $\frac{d}{f}$ rational curve $C_{\frac{d}{f}}$ in M_4^5. By picking up a fixed one-to-one holomorphic map $\phi_{0f} : CP^1 \to C_{\frac{d}{f}}$, we can write any holomorphic map in the component as $\phi_{0f} \circ m_f$, where $m_f : CP^1 \to CP^1$ is a degree f holomorphic map. Therefore, the connected component is identified with the moduli space $\mathcal{M}_{CP^1}(CP^1, f)$ of complex dimension $2f + 1$. On the other hand, $m_f \circ m_1$, the composition of m_f and $m_1 \in PSL(2, C)$, is also a degree f holomorphic map from CP^1 to CP^1. This means that the group $PSL(2, C)$ acts on $\mathcal{M}_{CP^1}(CP^1, f)$, which enables us to consider the quotient space $\mathcal{M}_{CP^1}(CP^1, f)/PSL(2, C)$. Its complex dimension is given by $2f + 1 - 3 = 2f - 2$, which equals the number of ψ zero modes. This quotient space is nothing but the moduli space $\mathcal{M}^R_{CP^1}(M_4^5, d)$ in the connected component, which was used in (3.48). We fix this degrees of freedom and write m_f as $m_f^0 \circ m_1$ (with m_f^0 fixed). Then we count the number of holomorphic maps that satisfy,

$$\phi_0(z_i) \in PD(h_i), \quad (i = 1, 2, 3), \tag{3.72}$$

by varying m_1. In this case, the image of ϕ_0, which we denote by $C_{\frac{d}{f}}$, is a degree $\frac{d}{f}$ rational curve and it intersects with H_i in $\frac{d}{f}$ points. But the number of inverse images of each intersection point is given by f since m_f^0 is a degree f map. Therefore, the number of points in CP^1 that are mapped to $PD(h_i)$ by $\phi_{0f} \circ m_f^0$ equals $\frac{d}{f} \cdot f = d$. This conclusion is the same as the one for type (a) components. Hence the number of holomorphic maps obtained from varying only m_1 is given by d^3. What remains to be done is to integrate out the degree of freedom of m_f^0 in the quotient space $\mathcal{M}_{CP^1}(CP^1, f)/PSL(2, C)$. According to (3.48), this contribution is evaluated by the formula

$$d^3 \int_{\mathcal{M}_{CP^1}(CP^1, f)/PSL(2,C)} c_{2f-2}(\mathcal{H}^1). \tag{3.73}$$

We omit details of evaluation of the above integral $\int_{\mathcal{M}_{CP^1}(CP^1, f)/PSL(2,C)} c_{2f-2}(\mathcal{H}^1)$ in this subsection because it is technically difficult. In conclusion, its value is given by $\frac{1}{f^3}$ [3]. Hence the contribution from a connected component of type (b) to $\langle \mathcal{O}_h(z_1)\mathcal{O}_h(z_2)\mathcal{O}_h(z_3)\rangle_d$ becomes $(\frac{d}{f})^3$. Since the number of connected components of type (b) is given by $n_{\frac{d}{f}}$, the number of degree $\frac{d}{f}$ rational curves, the discussions so far enable us to write down the following representation of the degree $d(\geq 1)$ correlation function $\langle \mathcal{O}_h(z_1)\mathcal{O}_h(z_2)\mathcal{O}_h(z_3)\rangle_d$:

$$\langle \mathcal{O}_h(z_1)\mathcal{O}_h(z_2)\mathcal{O}_h(z_3)\rangle_d = \sum_{f|d} (\frac{d}{f})^3 n_{\frac{d}{f}} = \sum_{f|d} (f)^3 n_f. \tag{3.74}$$

In the above formula, f can take the value d.

Lastly, we discuss the $d = 0$ case. A degree 0 holomorphic map from CP^1 to $M_4^5 \subset CP^4$ is given by a constant map,

$$\phi_0(s:t) = (a_0^1 : a_0^2 : a_0^3 : a_0^4 : a_0^5) \in M_4^5, \tag{3.75}$$

as can be seen from (3.52). Hence the moduli space $\mathcal{M}_{CP^1}(M_4^5, 0)$ becomes M_4^5 itself. Moreover, no ψ zero modes appear since the complex dimension of the moduli space $(= 3)$ equals the virtual dimension given by (3.67). Therefore, $\langle \mathcal{O}_h(z_1)\mathcal{O}_h(z_2)\mathcal{O}_h(z_3)\rangle_0$ equals the number of constant maps that satisfy (3.70), i.e., the number of points in $PD(h_1) \cap PD(h_2) \cap PD(h_3)$. If we pick up hyperplanes $X_i = 0$ $(i = 1, 2, 3)$ as representatives of $PD(h_i)$'s, $PD(h_1) \cap PD(h_2) \cap PD(h_3)$ is given by,

$$\begin{aligned} &\{(X_1 : X_2 : \cdots : X_5) \in CP^4 \mid X_1 = 0, X_2 = 0, X_3 = 0, \sum_{j=1} (X_j)^5 = 0\} \\ &= \{(0:0:0:1:e^{\frac{2\pi\sqrt{-1}j}{5}}) \in M_4^5 \subset CP^4 \mid j = 0, 1, \ldots, 4\}. \end{aligned} \tag{3.76}$$

Hence we obtain

$$\langle \mathcal{O}_h(z_1)\mathcal{O}_h(z_2)\mathcal{O}_h(z_3)\rangle_0 = 5. \tag{3.77}$$

Alternatively, this value is also obtained from

$$\int_{M_4^5} h \wedge h \wedge h = \int_{CP^4} 5h \cdot h^3 = 5 \int_{CP^4} h^4 = 5, \tag{3.78}$$

by turning submanifolds back in cohomology elements.

As was discussed in Sect. 3.2.1, the correlation function $\langle \mathcal{O}_h(z_1)\mathcal{O}_h(z_2)\mathcal{O}_h(z_3)\rangle$ of the topological sigma model is given by

$$\langle \mathcal{O}_h(z_1)\mathcal{O}_h(z_2)\mathcal{O}_h(z_3)\rangle = \sum_{d=0}^{\infty} \langle \mathcal{O}_h(z_1)\mathcal{O}_h(z_2)\mathcal{O}_h(z_3)\rangle_d e^{2\pi i d t}. \tag{3.79}$$

Plugging (3.77) and (3.74) into the above formula results in

$$\langle \mathcal{O}_h(z_1)\mathcal{O}_h(z_2)\mathcal{O}_h(z_3)\rangle = 5 + \sum_{d=1}^{\infty} \Big(\sum_{f|d} f^3 n_f\Big) e^{2\pi i d t}. \tag{3.80}$$

We then change the order of summation by using the relation: $f|d \Leftrightarrow d = mf$ (m: positive integer) and obtain the following formula:

$$\begin{aligned} \langle \mathcal{O}_h(z_1)\mathcal{O}_h(z_2)\mathcal{O}_h(z_3)\rangle &= 5 + \sum_{f=1}^{\infty}\sum_{m=1}^{\infty} f^3 n_f e^{2\pi i m f t} \\ &= 5 + \sum_{f=1}^{\infty} \frac{f^3 n_f e^{2\pi i f t}}{1 - e^{2\pi i f t}}, \end{aligned} \tag{3.81}$$

This formula coincides with the expansion form (1.28) of $\lambda_{ttt}(t)$, the Yukawa coupling associated with the Kähler deformation of M_4^5, which was used to count the number of degree d rational curves in M_4^5. In this way, we conclude that $\lambda_{ttt}(t)$ is identified with the correlation function $\langle \mathcal{O}_h(z_1)\mathcal{O}_h(z_2)\mathcal{O}_h(z_3)\rangle$ of the topological sigma model (A-model).

3.3 Topological Sigma Model (B-Model)

3.3.1 Lagrangian

In this section, we discuss the topological sigma model (B-model), which is obtained from applying the B-twist in Sect. 3.1 to the Lagrangian (3.1) of the $N = 2$ supersymmetric sigma model. As was introduced in (3.5), the B-twist is the operation to change the spin quantum number of the fields in the following way:

$$\phi^i, \phi^{\bar{i}} \to \phi^i, \phi^{\bar{i}}, \quad \psi_+^i \to \psi_z^i, \; \psi_-^i \to \psi_{\bar{z}}^i, \; \psi_+^{\bar{i}} \to \psi^{\bar{i}}, \; \psi_-^{\bar{i}} \to \psi^{\bar{i}}, \tag{3.82}$$

where we denote the fields obtained from $\psi_+^{\bar{i}}$ and $\psi_-^{\bar{i}}$ by $\psi_1^{\bar{i}}$ and $\psi_2^{\bar{i}}$ respectively. In constructing the Lagrangian of the B-model, we rename these fields by taking the linear combination of them:

$$\eta^{\bar{i}} := \psi_1^{\bar{i}} + \psi_2^{\bar{i}}, \quad \theta_i := g_{i\bar{j}}(\psi_1^{\bar{j}} - \psi_2^{\bar{j}}), \quad \rho_z^i := \psi_z^i, \quad \rho_{\bar{z}}^i := \psi_{\bar{z}}^i. \tag{3.83}$$

The BRST-transformation of the model is obtained by setting $\alpha_+ = \alpha_- = 0$, $\tilde{\alpha}_+ = \tilde{\alpha}_- = \alpha$ in (3.3):

$$\delta\phi^i = 0, \quad \delta\phi^{\bar{i}} = i\alpha\eta^{\bar{i}}, \quad \delta\eta^{\bar{i}} = \delta\theta_i = 0, \quad \delta\rho_z^i = -\alpha\partial_z\phi^i, \quad \delta\rho_{\bar{z}}^i = -\alpha\partial_{\bar{z}}\phi^i. \tag{3.84}$$

The BRST charge Q is defined by the relation $\delta X = i\alpha\{Q, X\}$ in the same way as the A-model, and it satisfies $Q^2 = 0$. The Lagrangian is given by

$$\begin{aligned} L = t\int_\Sigma d^2z\Big(& g_{i\bar{j}}(\partial_z\phi^i\partial_{\bar{z}}\phi^{\bar{j}} + \partial_z\phi^{\bar{j}}\partial_{\bar{z}}\phi^i) + i\eta^{\bar{j}}(D_z\rho_{\bar{z}}^i + D_{\bar{z}}\rho_z^i)g_{i\bar{j}} \\ & + i\theta_i(D_{\bar{z}}\rho_z^i - D_z\rho_{\bar{z}}^i) + R_{i\bar{j}k\bar{l}}\rho_z^i\rho_{\bar{z}}^k\eta^{\bar{j}}\theta_m g^{m\bar{l}}\Big). \end{aligned} \tag{3.85}$$

It is also written as follows:

$$L = it\int_\Sigma \{Q, V\} + tW, \tag{3.86}$$

where V and W are given by

$$V = g_{i\bar{j}}(\rho_z^i\partial_{\bar{z}}\phi^{\bar{j}} + \rho_{\bar{z}}^i\partial_z\phi^{\bar{j}}), \quad W = \int_\Sigma d^2z\big(i\theta_i(D_{\bar{z}}\rho_z^i - D_z\rho_{\bar{z}}^i) + R_{i\bar{j}k\bar{l}}\rho_z^i\rho_{\bar{z}}^k\eta^{\bar{j}}\theta_m g^{m\bar{l}}\big). \tag{3.87}$$

We can further rewrite W by using the differential form $\rho^i := \rho_z^i dz + \rho_{\bar{z}}^i d\bar{z}$, exterior differential operator $D := dz \wedge D_z + d\bar{z} \wedge D_{\bar{z}}$ with respect to the covariant derivative, and the relation $d^2z = -idz \wedge d\bar{z}$:

$$W = \int_\Sigma \Big(-\theta_i D\rho^i - \frac{i}{2}R_{i\bar{j}k\bar{l}}\rho^i \wedge \rho^k\eta^{\bar{j}}\theta_m g^{m\bar{l}}\Big). \tag{3.88}$$

In order to see that this theory is topological, we have to check whether it is invariant under deformation of metrics of the Riemann surface Σ and target space M. As for the term $it\int_\Sigma\{Q, V\}$ in (3.86), its response under deformation of metrics is written in the form $\{Q, *\}$. This kind of contribution to the path integral can be shown to vanish in the same way as the discussion of the A-model. The term W is written as the 2-form of Riemann surface Σ and it is invariant under deformation of the metric of Σ. On the other hand, it can be shown by elaborate computation that the response of W under deformation of metric of M is also written in the form $\{Q, *\}$. In this way, we can confirm that the B-model is also a topological field theory. As in the case of the A-model, correlation function of the B-model does not depend on the coupling constant t. This is clear with respect to the term $it \int_\Sigma \{Q, V\}$ since its response under deformation of t is written in the form $\{Q, *\}$. Moreover, we can make the term W independent of t by redefining $\theta_i \to \frac{\theta_i}{t}$. This operation is possible since W is linear in θ^i.

In contrast with the fact that the correlation function of the A-model is written as a power series of $e^{2\pi it}$ (only the degree d correlation function is t-independent), the correlation function of the B-model is entirely independent of t. Therefore, in order to compute correlation functions, we can take the weak coupling limit $t \to \infty$ and apply the saddle point approximation from the start. Saddle points, or solutions of the classical equation of motion of ϕ, are evaluated from L as follows:

$$\partial_z \phi^i = \partial_{\bar{z}} \phi^i = \partial_z \phi^{\bar{i}} = \partial_{\bar{z}} \phi^{\bar{i}} = 0. \tag{3.89}$$

In other words, they are given by constant maps. Solutions of the equation of motion of fermions, or fermion zero modes, correspond to the tangent space of the moduli space of solutions of ϕ. This moduli space is nothing but M, and numbers of zero modes of θ and η equal $\dim_{\mathbf{C}}(M)(= \dim_{\mathbf{C}}(T'M) = \dim_{\mathbf{C}}(\overline{T'}M))$ respectively. On the other hand, we have no ρ zero modes. We then expand the Lagrangian around saddle points up to second order and execute Gaussian integration. But the situation is more complicated than the case of the A-model. Since subscripts of ρ^i_z and $\rho^i_{\bar{z}}$ with respect to M are both holomorphic, the B-model is not symmetric under the interchange of holomorphic and anti-holomorphic degrees of freedom. This is in contrast with the A-model, which is symmetric with respect to these degrees of freedom. It causes the appearance of the anomaly in the process of Gaussian integration of fermion fields. This anomaly turns out to be proportional to $c_1(M)$, the first Chern class of M. **Therefore, the B-model is well-defined only if M is a Calabi–Yau manifold.** This characteristic is also in contrast with the A-model, which is well-defined if M is a Kähler manifold. Let us assume that M is a Calabi–Yau manifold and go into Gaussian integration. As one can see from the form of the BRST transformation, the B-model does not have symmetry between bosonic and fermionic degrees of freedom. Hence we cannot expect that the determinant of the differential operator of bosons and that of fermions cancel each other. But the non-trivial contribution obtained after Gaussian integration does not depend on position in M, and it can be absorbed by renormalization of the coupling constant t. Since the correlation function of the B-model is independent of t, we can ignore the non-trivial contribution of Gaussian integration. As can be seen from the discussions so far, the process of saddle point approximation is more complicated than the case of the A-model. But in conclusion, evaluation of the correlation function reduces to classical integration on M, which is the moduli space of solutions of the classical equation of motion. Hence the remaining process is simpler than that of the A-model.

3.3.2 Observable that Satisfies $\{Q, \mathcal{O}\} = 0$ and Topological Selection Rule

As was discussed in the case of the A-model, the observable $\mathcal{O}$ of the B-model is also required to satisfy the relation $\{Q, \mathcal{O}\} = 0$, i.e., to be invariant under BRST transformation. In order to find the observable of the B-model, we consider the degree $(0, p)$ differential form V on a Calabi–Yau manifold M which takes a value in the holomorphic vector bundle $\wedge^q T'M$ (a vector bundle obtained from taking exterior powers of $T'M$):

$$V = dx^{\bar{i}_1} \wedge dx^{\bar{i}_2} \wedge \cdots \wedge dx^{\bar{i}_p} V^{j_1 j_2 \cdots j_q}_{\bar{i}_1 \bar{i}_2 \cdots \bar{i}_p}(x) \partial_{j_1} \wedge \partial_{j_2} \wedge \cdots \partial_{j_q}. \tag{3.90}$$

Next, we introduce an observable $\mathcal{O}_V(z)$ associated with V:

$$\mathcal{O}_V(z) = \eta^{\bar{i}_1} \eta^{\bar{i}_2} \cdots \eta^{\bar{i}_p} V^{j_1 j_2 \cdots j_q}_{\bar{i}_1 \bar{i}_2 \cdots \bar{i}_p}(\phi(z)) \theta_{j_1} \theta_{j_2} \cdots \theta_{j_q}. \tag{3.91}$$

Then the following relation is derived from the BRST transformation (3.84):

$$\{Q, \mathcal{O}_V(z)\} = \mathcal{O}_{\overline{\partial} V}(z), \tag{3.92}$$

where $\overline{\partial}$ is anti-holomorphic exterior differential operator that was used to define Dolbeault cohomology. Its action on V is defined by

$$\overline{\partial} V = dx^{\bar{i}} \wedge dx^{\bar{i}_1} \wedge dx^{\bar{i}_2} \wedge \cdots \wedge dx^{\bar{i}_p} \left(\partial_{\bar{i}} V^{j_1 j_2 \cdots j_q}_{\bar{i}_1 \bar{i}_2 \cdots \bar{i}_p}(x)\right) \partial_{j_1} \wedge \partial_{j_2} \wedge \cdots \partial_{j_q}. \tag{3.93}$$

Then $\mathcal{O}_V(z)$ becomes an observable if V is $\overline{\partial}$ closed. If V is written as $V = \overline{\partial} W$, the relation $\mathcal{O}_V(z) = \{Q, \mathcal{O}_W(z)\}$ holds and the correlation function that contains $\mathcal{O}_V(z)$ vanishes. Hence the physical observables are in one-to-one correspondence with the elements of Dolbeault cohomology $H^{0,p}(M, \wedge^q T'M)$. Next, let us consider the correlation function:

$$\langle \mathcal{O}_{V_1}(z_1) \mathcal{O}_{V_2}(z_2) \cdots \mathcal{O}_{V_n}(z_n) \rangle = \int \mathcal{D}\phi \mathcal{D}\eta \mathcal{D}\theta \mathcal{D}\rho \exp(-L) \mathcal{O}_{V_1}(z_1) \mathcal{O}_{V_2}(z_2) \cdots \mathcal{O}_{V_n}(z_n). \tag{3.94}$$

If we apply saddle point approximation to the r.h.s., the remaining degrees of ϕ are just integration on M. As for fermionic degrees of freedom, there remains Grassmann integration of zero modes of η and θ, each of whose number is given by $n := \dim_{\mathbb{C}}(M)$. At this stage, integration of ρ is already done since there exist no ρ zero modes. On the other hand, the observable $\mathcal{O}_V(z)$ loses its dependence on position z on the Riemann surface Σ because ϕ becomes a constant map after saddle point approximation. Hence it turns into

$$\mathcal{O}_V(z) = \mathcal{O}_V = \eta_0^{\bar{i}_1} \eta_0^{\bar{i}_2} \cdots \eta_0^{\bar{i}_p} V^{j_1 j_2 \cdots j_q}_{\bar{i}_1 \bar{i}_2 \cdots \bar{i}_p}(x) \theta_{0,j_1} \theta_{0,j_2} \cdots \theta_{0,j_q}. \tag{3.95}$$

With these considerations, we rewrite (3.94) as follows:

$$\langle \mathcal{O}_{V_1}\mathcal{O}_{V_2}\cdots\mathcal{O}_{V_n}\rangle = \int_M d\phi_0 \int d\eta_0^{\bar{1}}\cdots d\eta_0^{\bar{n}} \int d\theta_{0,1}\cdots d\theta_{0,n}\mathcal{O}_{V_1}\mathcal{O}_{V_2}\cdots\mathcal{O}_{V_n}. \quad (3.96)$$

Let $\mathcal{O}_{V_I}$ be an observable constructed from $V_I \in H^{p_I}(M, \wedge^{q_I}T'M)$. Then $\mathcal{O}_{V_1}\mathcal{O}_{V_2}\cdots\mathcal{O}_{V_n}$ in the r.h.s. contains $\sum_{I=1}^n p_I$ η zero modes and $\sum_{I=1}^n q_I$ θ zero modes. Hence the relations

$$\sum_{I=1}^n p_I = n, \quad \sum_{I=1}^n q_I = n, \quad (3.97)$$

must hold to obtain a non-vanishing correlation function. The condition given by (3.97) is called the topological selection rule of the B-model.

3.3.3 Correlation Function

As can be seen from the discussion of the previous subsection, the correlation function of the B-model is evaluated from the formula:

$$\langle \mathcal{O}_{V_1}\mathcal{O}_{V_2}\cdots\mathcal{O}_{V_n}\rangle = \int_M d\phi_0 \int d\eta_0^{\bar{1}}\cdots d\eta_0^{\bar{n}} \int d\theta_{0,1}\cdots d\theta_{0,n}\mathcal{O}_{V_1}\mathcal{O}_{V_2}\cdots\mathcal{O}_{V_n}.$$
$$(V_I \in H^{p_I}(M, \wedge^{q_I}T'M), \ \sum_{I=1}^n p_I = n, \ \sum_{I=1}^n q_I = n). \quad (3.98)$$

This formula has some inconvenient aspects for geometrical interpretation. By the correspondence between V and $\mathcal{O}_V$ given in (3.95), $\mathcal{O}_{V_1}\mathcal{O}_{V_2}\cdots\mathcal{O}_{V_n}$ can be replaced by $\mathcal{O}_{V_1\wedge V_2\wedge\cdots\wedge V_n}$ (where wedge product means taking the exterior product both in $dx^{\bar{i}}$ and ∂_i). From the topological selection rule, $V_1 \wedge V_2 \wedge \cdots \wedge V_n$ becomes an element of $H^{0,n}(M, \wedge^n T'M)$. Since the l.h.s. of (3.98) is a correlation function of topological field theory, it is expected to be given as an integral of closed (n, n) form on M, which is an element of the Dolbeault cohomology $H^{0,n}(M, \wedge^n(T'M)^*)$. Therefore, there occurs the requirement to translate the element $V_1 \wedge V_2 \wedge \cdots \wedge V_n$ of $H^{0,n}(M, \wedge^n T'M)$ into an element of $H^{0,n}(M, \wedge^n(T'M)^*)$. In order to construct the rule of translation, we consult representation of the correlation function in operator formalism.

$$\langle \mathcal{O}_{V_1}\mathcal{O}_{V_2}\cdots\mathcal{O}_{V_n}\rangle = \langle 0|\mathcal{O}_{V_1}\mathcal{O}_{V_2}\cdots\mathcal{O}_{V_n}|0\rangle \quad (3.99)$$

In this formula, the correlation function is represented as the inner product of the state obtained from applying the operator $\mathcal{O}_{V_1}\mathcal{O}_{V_2}\cdots\mathcal{O}_{V_n}$ onto the vacuum state $|0\rangle$ and the dual vacuum state $\langle 0|$. Then what is the vacuum state in the B-model? At this stage, we recall the fact that the B-model is well-defined only on a Calabi–Yau manifold M. When M is a Calabi–Yau manifold, its first Chern class $c_1(M)(= c_1(T'M))$ vanishes. Let us consider here the holomorphic line bundle $\wedge^n T'M$ on M (it is rank 1 since $\dim_{\mathbf{C}}(M) = n$). Its first Chern class is shown to equal $c_1(T'M)$ by using the splitting principle in Chap. 2, and it also vanishes. Therefore, $\wedge^n T'M$ is isomorphic to the trivial line bundle $M \times \mathbf{C}$ as a vector bundle on M. Then $\wedge^n (T'M)^*$, which is the dual vector bundle of $\wedge^n T'M$, is also isomorphic to $M \times \mathbf{C}$. We now focus on the Dolbeault cohomology $H^0(M, \wedge^n (T'M)^*)$. It is a vector space spanned by linearly independent global holomorphic sections of $\wedge^n (T'M)^*$. Since $\wedge^n (T'M)^*$ is a trivial line bundle, its global holomorphic sections consist only of constant sections. Hence we obtain $\dim_{\mathbf{C}}(H^0(M, \wedge^n (T'M)^*)) = 1$. This result tells us that there exists a holomorphic $(n, 0)$ form

$$\Omega = \Omega_{i_1 i_2 \cdots i_n} dx^{i_1} \wedge dx^{i_2} \wedge \cdots \wedge dx^{i_n}, \tag{3.100}$$

which is unique up to multiplication of constants. Putting together the fact that the B-model is well-defined only on a Calabi–Yau manifold and the fact that Ω is uniquely determined only for a Calabi–Yau manifold, we are inclined to propose the following correspondence:

$$|0\rangle \longleftrightarrow \Omega. \tag{3.101}$$

Under this proposal, we consider the action of $\mathcal{O}_{V_1}\mathcal{O}_{V_2}\cdots\mathcal{O}_{V_n} = \mathcal{O}_{V_1\wedge V_2 \wedge \cdots \wedge V_n}$ on the vacuum Ω. For this purpose, we define the natural exterior product of $V \in H^{0,p}(M, \wedge^q T'M)$ and $W \in H^{0,r}(M, \wedge^s (T'M)^*)(q \le s)$ as follows:

$$\begin{aligned} V \wedge W &= (dx^{\bar{i}_1} \wedge \cdots \wedge dx^{\bar{i}_p} V^{j_1 \cdots j_q}_{\bar{i}_1 \cdots \bar{i}_p}(x) \partial_{j_1} \wedge \cdots \partial_{j_q}) \\ &\quad \wedge (dx^{\bar{k}_1} \wedge \cdots \wedge dx^{\bar{k}_r} W_{\bar{k}_1 \cdots \bar{k}_r l_1 \cdots l_s}(x) dx^{l_1} \wedge \cdots \wedge dx^{l_s}) \\ &:= dx^{\bar{i}_1} \wedge \cdots \wedge dx^{\bar{i}_p} \wedge dx^{\bar{k}_1} \wedge \cdots \wedge dx^{\bar{k}_r} V^{j_1 \cdots j_q}_{\bar{i}_1 \cdots \bar{i}_p}(x) \\ &\quad \times W_{\bar{k}_1 \cdots \bar{k}_r j_1 \cdots j_q l_1 \cdots l_{s-q}}(x) dx^{l_1} \wedge \cdots \wedge dx^{l_{s-q}}. \end{aligned} \tag{3.102}$$

Under this definition, $V \wedge W$ becomes an element of $H^{0,p+r}(M, \wedge^{s-q}(T'M)^*)$. By using this exterior product, the following correspondence:

$$\begin{aligned} d\mathcal{O}_{V_1}\mathcal{O}_{V_2}\cdots\mathcal{O}_{V_n}|0\rangle \longleftrightarrow \;& (V_1 \wedge V_2 \wedge \cdots \wedge V_n) \wedge \Omega \\ &= V_1 \wedge (V_2 \wedge (\cdots \wedge (V_n \wedge \Omega)\cdots)) \\ &= V_1 \wedge V_2 \wedge \cdots \wedge V_n \wedge \Omega \end{aligned} \tag{3.103}$$

turns $\mathcal{O}_{V_1}\mathcal{O}_{V_2}\cdots\mathcal{O}_{V_n}|0\rangle$ into a $(0, n)$ form on M. Lastly, the correspondence

$$\text{taking the inner product with } \langle 0| \longleftrightarrow \int_M \Omega \wedge, \tag{3.104}$$

leads us to the following formula:

$$\langle \mathcal{O}_{V_1} \mathcal{O}_{V_2} \cdots \mathcal{O}_{V_n} \rangle = \langle 0 | \mathcal{O}_{V_1} \mathcal{O}_{V_2} \cdots \mathcal{O}_{V_n} | 0 \rangle = \int_M \Omega \wedge (V_1 \wedge V_2 \wedge \cdots \wedge V_n \wedge \Omega). \quad (3.105)$$

The r.h.s. becomes an integral of (n, n) form on M, which is a well-defined correlation function of topological field theory. If M is a complex 3-dimensional Calabi–Yau manifold X, we can consider the correlation function $\langle \mathcal{O}_{u_{(\alpha)}} \mathcal{O}_{u_{(\beta)}} \mathcal{O}_{u_{(\gamma)}} \rangle$ for $u_{(\alpha)}, u_{(\beta)}, u_{(\gamma)} \in H^1(X, T'X)$, which is given by

$$\langle \mathcal{O}_{u_{(\alpha)}} \mathcal{O}_{u_{(\beta)}} \mathcal{O}_{u_{(\gamma)}} \rangle = \int_X \Omega \wedge \left(u_{(\alpha)} \wedge u_{(\beta)} \wedge u_{(\gamma)} \wedge \Omega \right). \quad (3.106)$$

The r.h.s. coincides with the formula of the Yukawa coupling $\lambda_{\alpha\beta\gamma}$ associated with deformation of the complex structure of X, which was mentioned in Chap. 1. The discussions in this chapter conclude that coincidence between the Yukawa coupling associated with the Kähler deformation of the complex 3-dimensional Calabi–Yau manifold X and the Yukawa coupling associated with deformation of the complex structure of its mirror manifold X^* can be reinterpreted by the following correspondence:

$$\text{correlation function of the } A-\text{model on } X \longleftrightarrow \text{correlation function of the } B-\text{model on } X^*.$$

References

1. E. Witten. *Mirror Manifolds and Topological Field Theory*. Essays on Mirror Manifolds (ed. S.T. Yau), International Press, Hong Kong (1992)
2. K. Hori. *Constraints for topological strings in $D \geq 1$*. Nuclear Phys. B 439 (1995), no. 1-2, 395–420
3. M. Kontsevich. Enumeration of rational curves via torus actions. The Moduli Space of Curves (Texel Island, 1994), 335–368, Progr. Math., 129, Birkhäuser Boston, Boston, MA (1995)

Chapter 4
Details of B-Model Computation

Abstract In this chapter, we introduce toric geometry, which is useful to construct a pair of complex 3-dimensional Calabi–Yau manifolds that are mirrors of each other. First, we explain how to construct a toric manifold from a pair (Δ, Δ^*) of reflexive convex polytope Δ and its dual polytope Δ^*. A pair of mirror manifolds is constructed from the operation of exchanging roles of the polytope Δ and its dual Δ^*. Next, we evaluate the period integral of a complex 3-dimensional Calabi–Yau hypersurface in a complex 4-dimensional toric manifold by using its toric data [3]. Lastly, we compute the Yukawa coupling of the B-model on a complex 3-dimensional Calabi–Yau manifold and explain the process of translating it into the Yukawa coupling of the A-model on its mirror manifold. (For details of toric geometry, we recommend readers to consult [2].)

4.1 Toric Geometry

4.1.1 Outline of Toric Geometry

We start by introducing an n-dimensional lattice (free abelian group of rank n) $M = \mathbf{Z}^n$ and its dual lattice $N = \mathbf{Z}^n$. We denote n-dimensional Euclid spaces obtained from tensoring M and N with $\mathbf{R}$ by $M_{\mathbf{R}} := M \otimes_{\mathbf{Z}} \mathbf{R} = \mathbf{R}^n$ and $N_{\mathbf{R}} := N \otimes_{\mathbf{Z}} \mathbf{R} = \mathbf{R}^n$ respectively. We also denote the inner product of $u \in M_{\mathbf{R}}$ and $v \in N_{\mathbf{R}}$ by $\langle u, v\rangle$. This inner product satisfies $\langle e_i, f^j\rangle = \delta_i^j$ for the standard basis e_i, $(i = 1, 2, \ldots, n)$ of $M_{\mathbf{R}}$ and f^j $(j = 1, 2, \ldots, n)$ of $N_{\mathbf{R}}$.

With these settings, we consider an n-dimensional convex polytope Δ in $M_{\mathbf{R}}$ whose vertices are given by lattice points of M. We impose the following conditions on Δ.

- **Δ contains only 0 as lattice points except for its boundary points.**

Next, we define a polytope Δ^* in $N_{\mathbf{R}}$ which is dual to Δ by the following formula:

$$\Delta^* = \{v \in N_{\mathbf{R}} \mid \langle u, v\rangle \geq -1, \ \forall u \in \Delta\ \}. \tag{4.1}$$

M. Jinzenji, *Classical Mirror Symmetry*, SpringerBriefs in Mathematical Physics,
https://doi.org/10.1007/978-981-13-0056-1_4

At first sight, it seems difficult to extract geometrical information on Δ^* from this definition. We will discuss a practical way to obtain information on Δ^* by taking a specific example. Δ is called reflexive if Δ^* becomes a polytope whose vertices are given by lattice points of N. The reason why we use the word "reflexive" comes from the fact that for Δ^* whose vertices lie in N, $(\Delta^*)^*$ defined by

$$(\Delta^*)^* = \{u \in M_{\mathbf{R}} \mid \langle u, v\rangle \geq -1, \ \forall v \in \Delta^* \}, \tag{4.2}$$

turns out to be Δ. Hence Δ^* is also reflexive if Δ is reflexive.

To be brief, toric geometry is a recipe to construct a complex n-dimensional space (or orbifold) from information on a pair of polytopes (Δ, Δ^*).

4.1.2 An Example: Complex Projective Space

In this subsection, we construct a complex projective space CP^n by using the recipe of toric geometry. The reflexive polytope Δ used for construction of CP^n is a convex polytope in $M_{\mathbf{R}} = \mathbf{R}^n$ whose vertices are given by the following $(n+1)$ integral vectors in M:

$$\begin{aligned} u_1 &= (n, -1, -1, \ldots, -1), \\ u_2 &= (-1, n, -1, \ldots, -1), \\ &\vdots \\ u_n &= (-1, -1, \ldots, -1, n), \\ u_{n+1} &= (-1, -1, -1, \ldots, -1). \end{aligned} \tag{4.3}$$

This polytope is an n-dimensional $(n+1)$-gon in $\mathbf{R}^n$, which has a minimal number of $(n-1)$-faces as an n-dimensional polytope. In order to determine Δ^* explicitly, we introduce the following theorem without proof.

Theorem 4.1.1 *Let Δ be a reflexive polytope in $M_{\mathbf{R}} = \mathbf{R}^n$. We denote the $(n-1)$-faces of Δ by $\Delta^j_{(n-1)}$ $(j = 1, 2, \ldots, k)$. Then the defining equation of the $(n-1)$-plane that contains $\Delta^j_{(n-1)}$ is always written by using $v_j \in N$ as follows:*

$$\langle x, v_j\rangle = -1, \tag{4.4}$$

and Δ^* is given by the convex hull of v_j $(j = 1, 2, \ldots, k)$.

Of course, we also have the dual version of this theorem which is obtained by exchanging the roles of M and N.

Theorem 4.1.2 *Let Δ^* be a reflexive polytope in $N_{\mathbf{R}} = \mathbf{R}^n$ and $\Delta^{*,j}_{(n-1)}$ $(j = 1, 2, \ldots, l)$ be $(n-1)$-faces of Δ^*. Then the defining equation of an $(n-1)$-plane that contains $\Delta^{*,j}_{(n-1)}$ is always written by using $u_j \in M$ as follows:*

$$\langle u_j, y \rangle = -1, \tag{4.5}$$

and Δ is given by the convex hull of u_j ($j = 1, 2, \ldots, l$).

With the aid of these theorems, we can determine the vertices of Δ^* from defining equations of $(n-1)$-planes that contain $(n-1)$-faces of Δ. Let $\Delta^i_{(n-1)}$ ($i = 1, 2, \ldots, n+1$) be an $(n-1)$-face that is the convex hull of $u_1, \ldots, u_{i-1}, u_{i+1}, \ldots, u_{n+1}$. The defining equation of the $(n-1)$-plane that contains $\Delta^i_{(n-1)}$ is given as follows:

$$\begin{aligned} x_i &= -1 \quad (i = 1, \ldots, n), \\ -x_1 - x_2 - \cdots - x_n &= -1 \quad (i = n+1). \end{aligned} \tag{4.6}$$

Hence Δ^* is the convex hull of the following $(n+1)$ vertices in N:

$$\begin{aligned} v_1 &= (1, 0, 0, \ldots, 0), \\ v_2 &= (0, 1, 0, \ldots, 0), \\ &\vdots \\ v_n &= (0, 0, \ldots, 0, 1), \\ v_{n+1} &= (-1, -1, \ldots, -1). \end{aligned} \tag{4.7}$$

In this way, we have obtained (Δ, Δ^*), a pair of a polytope in $M_{\mathbf{R}}$ and one in $N_{\mathbf{R}}$. There remain several steps to construct the complex projective space CP^n from the pair. Let us introduce here a cone in the Euclidean space $\mathbf{R}^n$.

Definition 4.1.1 Let $w_1, w_2, \ldots, w_m$ be vectors in $\mathbf{R}^n$. A cone in $\mathbf{R}^n$ generated by these vectors is defined by

$$\sigma = \{\sum_{j=1}^{m} \lambda^j w_j \in \mathbf{R}^n \mid \lambda^j \geq 0, \ (j = 1, 2, \ldots, m)\}. \tag{4.8}$$

We also write it as $\sigma =< w_1, w_2, \ldots, w_m >_{\mathbf{R}_{\geq 0}}$ when we want to specify generators. We call σ a k-dimensional cone if the dimension of the vector space generated by $w_1, w_2, \ldots, w_m$ is given by k. We define the 0-dimensional cone by $\{\mathbf{0}\}$. If the generators $w_1, w_2, \ldots, w_m$ are linearly independent integral vectors and if any lattice point of σ (element of $\sigma \cap \mathbf{Z}^n$) is represented by an integral linear combination of generators, the cone σ is called non-singular. If a cone σ is not non-singular, we call it a singular cone.

Then we focus on Δ^*. It also has $(n+1)$ $(n-1)$-faces. Let us consider $(n+1)$ $(n-1)$-dimensional cones, each of which is generated by vertices of the $(n-1)$-face. They are explicitly written as follows:

$$\sigma_i = < v_1, \dots, v_{i-1}, v_{i+1}, \dots, v_n, v_{n+1} >_{\mathbf{R}_{\geq 0}} \quad (i = 1, \dots, n),$$
$$\sigma_{n+1} = < v_1, v_2, \dots, v_n >_{\mathbf{R}_{\geq 0}} . \tag{4.9}$$

σ_i $(i = 1, 2, \dots, n+1)$ are n-dimensional cones in $N_{\mathbf{R}}$ and are all non-singular. These cones satisfy the relation $\cup_{i=1}^{n+1} \sigma_i = N_{\mathbf{R}}$ and $\sigma_i \cap \sigma_j$ $(i \neq j)$ becomes a cone whose dimension is no more than $(n-1)$. We denote a set of cones such as $\{\sigma_i \mid i = 1, 2, \dots, n+1\}$, i.e., that defines the partition of $N_{\mathbf{R}}$, by a fan.

We introduce another notation.

Definition 4.1.2 For an n-dimensional cone σ in $N_{\mathbf{R}}$, we define a dual cone σ^* in $M_{\mathbf{R}}$ as follows:

$$\sigma^* := \{ u \in M_{\mathbf{R}} \mid \langle u, v \rangle \geq 0, \ \forall v \in \sigma \}. \tag{4.10}$$

Then the cone σ_i^* dual to σ_i in (4.9) is computed as follows:

$$\sigma_i^* = < e_1 - e_i, \dots, e_{i-1} - e_i, e_{i+1} - e_i, \dots, e_n - e_i, -e_i >_{\mathbf{R}_{\geq 0}}$$
$$(i = 1, \dots, n),$$
$$\sigma_{n+1}^* = < e_1, e_2, \dots, e_n >_{\mathbf{R}_{\geq 0}}, \tag{4.11}$$

where e_i $(i = 1, 2, \dots, n)$ are the standard basis of $M_{\mathbf{R}}$. In order to determine generators of σ_i^*, we remove a generator from n generators of σ_i and find a vector in M that is orthogonal to the remaining $(n-1)$ generators and has inner product 1 with the removed generator. From the dual cone σ_i^*, we can construct local coordinate system $U_{(i)} = \{(z_{(i)}^1, z_{(i)}^2, \dots, z_{(i)}^n)\} = \mathbf{C}^n$ of CP^n. Correspondence between the coordinate $z_{(i)}^j$ and the generator of σ_i^* is given as follows. If $1 \leq i \leq n$, it is given by

$$z_{(i)}^j \leftrightarrow e_j - e_i, \quad (j = 1, \dots, i-1),$$
$$z_{(i)}^j \leftrightarrow e_{j+1} - e_i, \quad (j = i, \dots, n-1),$$
$$z_{(i)}^n \leftrightarrow -e_i, \tag{4.12}$$

and if $i = n+1$, it is given by

$$z_{(n+1)}^j \leftrightarrow e_j, \quad (j = 1, 2, \dots, n). \tag{4.13}$$

In this correspondence, additive operations between generators are translated into product operations between local coordinates and this translation leads us to coordinate transformation. For example, let us derive the coordinate transformation between $U_{(1)}$ and $U_{(2)}$. $z_{(1)}^j$ $(1 \leq j \leq n-1)$ corresponds to $e_{j+1} - e_1$ and $z_{(1)}^n$ corresponds to $-e_1$. On the other hand, $z_{(2)}^1$ corresponds to $e_1 - e_2$, $z_{(2)}^j$ $(2 \leq j \leq n-1)$ corresponds to $e_{j+1} - e_2$, and $z_{(2)}^n$ corresponds to $-e^2$. Hence we can obtain the following coordinate transformation by representing the generator that corresponds to $z_{(2)}^j$ by a linear combination of generators that correspond to $z^j_{(1)}$:

$$e_1 - e_2 = -(e_2 - e_1) \Rightarrow z^1_{(2)} = \frac{1}{z^1_{(1)}},$$

$$e_{j+1} - e_2 = (e_{j+1} - e_1) - (e_2 - e_1) \Rightarrow z^j_{(2)} = \frac{z^j_{(1)}}{z^1_{(1)}}, \ (j = 2, \ldots, n-1),$$

$$-e_2 = (-e_1) - (e_2 - e_1) \Rightarrow z^n_{(2)} = \frac{z^n_{(1)}}{z^1_{(1)}}. \tag{4.14}$$

These results coincide with the coordinate transformation between $U_{(1)}$ and $U_{(2)}$ derived in (2.12). We leave the other cases as exercises for the readers. In this way, we can confirm that the complex manifold constructed from (Δ, Δ^*) given by (4.3) and (4.7) is the complex projective space CP^n. We can apply the same procedure to other pair (Δ, Δ^*) of reflexive polytopes and construct rule of coordinate transformation between local coordinate systems that correspond to dual cones in $M_{\mathbf{R}}$. This is the fundamental recipe to construct a complex n-dimensional space from (Δ, Δ^*), but it is difficult for us to grasp the global geometrical structure of the manifold from the local coordinate system and rules of coordinate transformation. In the case of CP^n, homogeneous coordinates X_i $(i = 1, 2, \ldots, n+1)$ are convenient for this purpose. Is there any method to construct homogeneous coordinates for the complex n-dimensional manifold obtained from a general (Δ, Δ^*)? Fortunately, there exists a method suitable for this purpose. Let us turn into demonstration of this method. First, we take CP^n as an example. Δ^* in (4.7) has $(n+1)$ vertices and we associate a homogeneous coordinate X_i with each vertex v_i. For a general Δ^* whose vertices are given by $\{v_1, v_2, \ldots, v_{n+l}\}$, we also associate homogeneous coordinate X_i $(i = 1, 2, \ldots, n+l)$ with each vertex v_i. Next, we consider linear relations between v_i's. In the CP^n case, we have one non-trivial linear relation because v_i's are $(n+1)$ vectors in n-dimensional Euclidean space:

$$v_1 + v_2 + v_3 + \ldots + v_{n+1} = \mathbf{0}. \tag{4.15}$$

If we write the above relation in the following form:

$$\begin{pmatrix} 1 & 1 & 1 & \cdots & 1 \end{pmatrix} \begin{pmatrix} v_1 \\ v_2 \\ v_3 \\ \vdots \\ v_{n+1} \end{pmatrix} = \mathbf{0}, \tag{4.16}$$

we can read off the equivalence relation between homogeneous coordinates,

$$(X_1, X_2, \ldots, X_{n+1}) \sim (\lambda X_1, \lambda X_2, \ldots, \lambda X_{n+1}) \ (\lambda \in \mathbf{C}^{\times}), \tag{4.17}$$

from the row vector $\begin{pmatrix} 1 & 1 & 1 & \cdots & 1 \end{pmatrix}$. What we are saying is that the powers of λ multiplied by the X_i's are all given by 1. This gives us the equivalence relation to define CP^n by using homogeneous coordinates.

In the case of a general Δ^*, we have l linear relations,

$$\alpha_i^1 v_1 + \alpha_i^2 v_2 + \cdots + \alpha_i^{n+l} v_{n+l} = 0 \ (i = 1, 2, \ldots, l), \tag{4.18}$$

and we impose the following l equivalence relations on homogeneous coordinates:

$$\begin{aligned}
&(X_1, X_2, \ldots, X_{n+l}) \\
&\sim ((\lambda_1)^{\alpha_1^1} X_1, (\lambda_1)^{\alpha_1^2} X_2, \ldots, (\lambda_1)^{\alpha_1^{n+l}} X_{n+l}), \quad (\lambda_1 \in \mathbf{C}^\times), \\
&\sim ((\lambda_2)^{\alpha_2^1} X_1, (\lambda_2)^{\alpha_2^2} X_2, \ldots, (\lambda_2)^{\alpha_2^{n+l}} X_{n+l}), \quad (\lambda_2 \in \mathbf{C}^\times), \\
&\quad \vdots \\
&\sim ((\lambda_l)^{\alpha_l^1} X_1, (\lambda_l)^{\alpha_l^2} X_2, \ldots, (\lambda_l)^{\alpha_l^{n+l}} X_{n+l}), \quad (\lambda_l \in \mathbf{C}^\times).
\end{aligned} \tag{4.19}$$

By the way, in the CP^n case, we have to remove $\mathbf{0}$ from $\mathbf{C}^{n+1} = \{(X_1, X_2, \ldots, X_{n+1})\}$ in applying the equivalence relation (4.17). How can we determine the set to remove from $\mathbf{C}^{n+l} = \{(X_1, X_2, \ldots, X_{n+l})\}$ in the case of a general Δ^*? For this purpose, let us introduce the following definition.

Definition 4.1.3 Let $V(\Delta^*) := \{v_1, v_2, \ldots, v_{n+l}\}$ be the set of vertices of the n-dimensional convex polytope Δ^*. We call a subset $S = \{v_{i_1}, v_{i_2}, \ldots, v_{i_m}\}$ $(1 \leq i_1 \leq i_2 \leq \ldots \leq i_m \leq n + l)$ of $V(\Delta^*)$, which satisfies the following conditions, a primitive set:

(i) All the vertices in S do not belong to the same $(n-1)$-face of Δ^*.
(ii) Vertices that belong to a proper subset of S are always contained in an $(n-1)$-face of Δ^*.

Note that the number of primitive sets of Δ^* is not always one, though the primitive set of Δ^* given by (4.7) (CP^n case) is given by $\{v_1, v_2, \ldots, v_{n+1}\}$. Then we define a subset $E(S)$ of $\mathbf{C}^{n+l}$ by

$$E(S) := \{(X_1, X_2, \ldots, X_{n+l}) \in \mathbf{C}^{n+l} \mid X_{i_j} = 0, \ (j = 1, 2, \ldots, m)\}. \tag{4.20}$$

Let $S_1, S_2, \ldots, S_h$ be primitive sets of Δ^*. Then the set to remove from $\mathbf{C}^{n+l}$ in imposing the equivalence relation (4.19) is given as follows:

$$Z(\Delta^*) = \bigcup_{j=1}^{h} E(S_j). \tag{4.21}$$

From this recipe, we can confirm that the set to remove from $\mathbf{C}^{n+1}$ in constructing CP^n is given by $\{\mathbf{0}\}$.

Let us summarize the construction by using homogeneous coordinates:

(i) We first associate the homogeneous coordinate X_i with each vertex v_i ($i = 1, 2, \ldots, n+l$) of Δ^*.
(ii) We then remove $Z(\Delta^*)$ given by (4.21) from $\mathbf{C}^{n+l} = \{(X_1, X_2, \ldots, X_{n+l})\}$.
(iii) We impose l $\mathbf{C}^\times$-equivalence relations given by (4.19) on $\mathbf{C}^{n+l} - Z(\Delta^*)$.

The resulting quotient space $(\mathbf{C}^{n+l} - Z(\Delta^*))/(\mathbf{C}^\times)^l$ is the description of the complex n-dimensional space obtained from (Δ, Δ^*) by homogeneous coordinates. Note here that there exist some subtleties in this homogeneous coordinate description. As we have mentioned before, we can construct a fan, which gives us the partition of $N_\mathbf{R}$ by n-dimensional cones, from Δ^*. If all the n-dimensional cones in the fan are non-singular, the complex n-dimensional space becomes a manifold. In this case, the above description provides us all the geometrical information on the manifold. But if the fan contains n-dimensional cones that are singular, the complex n-dimensional space becomes an orbifold. Then the above description turns out to be a coarse description, which does not provide us with the detailed structure of orbifold singularities. In such cases, we have to investigate lattice points of singular cones to obtain detailed information on orbifold singularities. We leave the discussions on singular cases to the next section.

Before closing this subsection, we discuss another example with $n = 2$, where a reflexive polytope Δ is given by the convex hull of the following four vertices in $M_\mathbf{R} = \mathbf{R}^2$:

$$u_1 = (1, 1), \ u_2 = (-1, -1), \ u_3 = (1, -1), \ u_4 = (-1, 1). \tag{4.22}$$

Then Δ^* turns out to be the convex hull of the following four vertices in $N_\mathbf{R} = \mathbf{R}^2$:

$$v_1 = (1, 0), \ v_2 = (-1, 0), \ v_3 = (0, 1), \ v_4 = (0, -1). \tag{4.23}$$

Following the recipe of homogeneous coordinate description, we associate the homogeneous coordinate X_i with the vertex v_i. Let $Z(\Delta^*)$ be the set to remove from $\mathbf{C}^4 = \{(X_1, X_2, X_3, X_4)\}$ in imposing the $\mathbf{C}^\times$ equivalence relation. We determine primitive sets of Δ^* to obtain $Z(\Delta^*)$. Pairs of vertices which span four 1-faces of Δ^* are given by (v_1, v_3), (v_1, v_4), (v_2, v_3) and (v_2, v_4). The number of vertices of primitive set S is less than four because three different vertices cannot lie on the same 1-face. Then let us assume that S consists of three vertices. In this case, we can set $S = \{v_1, v_2, v_3\}$ without loss of generality. But its proper subset $\{v_1, v_2\}$ is not contained in a 1-face of Δ^*. Therefore, S consists of less than three vertices. Obviously, a set that consists of one vertex does not satisfy the condition of a primitive subset. Hence S consists of two vertices. Since the two vertices must not belong to the same 1-face, there exist two primitive sets $\{v_1, v_2\}$ and $\{v_3, v_4\}$. In conclusion, $Z(\Delta^*)$ is given as follows:

$$Z(\Delta^*) = \{(0, 0, X_3, X_4)\} \cup \{(X_1, X_2, 0, 0)\}. \tag{4.24}$$

Hence we obtain

$$\mathbf{C}^4 - Z(\Delta^*) = \{(X_1, X_2, X_3, X_4) \mid (X_1, X_2) \neq (0,0),\ (X_3, X_4) \neq (0,0)\}. \quad (4.25)$$

On the other hand, the four vectors v_1, v_2, v_3 and v_4 in $N_{\mathbf{R}} = \mathbf{R}^2$ satisfy two linear relations, and they are given by

$$v_1 + v_2 = 0, \quad v_3 + v_4 = 0. \quad (4.26)$$

Therefore, the equivalence relation imposed on $\mathbf{C}^4 - Z(\Delta^*)$ becomes,

$$\begin{aligned} &(X_1, X_2, X_3, X_4) \\ &\sim (\lambda_1 X_1, \lambda_1 X_2, X_3, X_4) \ \ (\lambda_1 \in \mathbf{C}^\times), \\ &\sim (X_1, X_2, \lambda_2 X_3, \lambda_2 X_4) \ \ (\lambda_2 \in \mathbf{C}^\times). \end{aligned} \quad (4.27)$$

Since the 2-dimensional cones associated with Δ^* are all non-singular, the description by homogeneous coordinates is complete. The results obtained so far tell us that the 2-dimensional complex manifold obtained from the pair (Δ, Δ^*) is nothing but $CP^1 \times CP^1$.

4.2 Formulation of Mirror Symmetry by Toric Geometry

4.2.1 An Example: Quintic Hypersurface in CP^4

In this subsection, we construct a pair of complex 3-dimensional Calabi–Yau manifolds that are mirrors of each other, by using toric geometry.

First, we pick up a quintic hypersurface in CP^4 and its mirror manifold as an example. Let us remind the readers of the pair of reflexive convex polytopes (Δ, Δ^*) that was used for construction of CP^4. Δ is the convex hull of the following vertices in $M_{\mathbf{R}} = \mathbf{R}^4$:

$$\begin{aligned} &u_1 = (4, -1, -1, -1), u_2 = (-1, 4, -1, -1), u_3 = (-1, -1, 4, -1), \\ &u_4 = (-1, -1, -1, 4), u_5 = (-1, -1, -1, -1), \end{aligned} \quad (4.28)$$

and Δ^* is the convex hull of the five vertices in $N_{\mathbf{R}} = \mathbf{R}^4$ that are given by

$$\begin{aligned} &v_1 = (1, 0, 0, 0), v_2 = (0, 1, 0, 0), v_3 = (0, 0, 1, 0), v_4 = (0, 0, 0, 1), \\ &v_5 = (-1, -1, -1, -1). \end{aligned} \quad (4.29)$$

As can be noticed from the discussions in the previous section, substantial information used for construction of CP^4 was obtained from Δ^*. But Δ plays important roles in constructing the Calabi–Yau manifold in CP^4, i.e., the quintic hypersurface. In general, the defining equation of the Calabi–Yau hypersurface in the complex 4-dimensional manifold (or orbifold) constructed from the pair (Δ, Δ^*) is written down by using data on lattice points of Δ (or points in $\Delta \cap M$). We identify M with $\mathbf{Z}^n$ and denote points in $M \cap \Delta$ by $\nu_i = (\nu_i^1, \nu_i^2, \ldots, \nu_i^n)$. Next, we associate the standard basis e_i $(i = 1, 2, \ldots, n)$ of M with variables t_i $(i = 1, 2, \ldots, n)$ which take values in $\mathbf{C}^\times$. We also introduce the notation $t^\nu := \prod_{j=1}^n t_j^{\nu^j}$. With this set-up, a defining equation of the Calabi–Yau hypersurface is given by

$$f_\Delta(t_1, t_2, \ldots, t_n) = \sum_{\nu_i \in M \cap \Delta} a_{\nu_i} t^{\nu_i} = 0, \tag{4.30}$$

where a_{ν_i} is a complex parameter associated with ν_i. But we have to make it clear how t_i's are related to homogeneous coordinates of the complex space constructed from the pair (Δ, Δ^*). For this purpose, we explain a recipe to obtain the defining equation of a quintic hypersurface in CP^4 from (4.30). First, we consider the local coordinate system. In order to construct the local coordinate system, we have to pick up a 4-dimensional cone σ in the fan obtained from Δ^*. For simplicity, we take the cone $\sigma =< \nu_1, \nu_2, \nu_3, \nu_4 >_{\mathbf{R}_{\geq 0}}$. Then we determine a cone $\sigma^* \subset M_{\mathbf{R}}$ which is dual to σ and associate local coordinate variables with the basis of σ^*. We can easily see that σ^* is given by $< e_1, e_2, e_3, e_4 >_{\mathbf{R}_{\geq 0}}$. Hence we associate the variable z_i with e_i and obtain the local coordinate system (z_1, z_2, z_3, z_4). On the other hand, t_i also corresponds to e_i and we obtain the relation $t_i = z_i$ in this case. From the discussions on construction of the local coordinate system of CP^n in the previous section, we can observe that these local coordinates are written in terms of homogeneous coordinates as follows:

$$(X_1 : X_2 : X_3 : X_4 : X_5) = \left(\frac{X_1}{X_5} : \frac{X_2}{X_5} : \frac{X_3}{X_5} : \frac{X_4}{X_5} : 1\right) = (z_1 : z_2 : z_3 : z_4 : 1). \tag{4.31}$$

Hence we obtain $t_i = \frac{X_i}{X_5}$. We plug this relation into f_δ and multiply it by $X_1X_2X_3X_4X_5$ to make the result into a homogeneous polynomial. Then we obtain the defining equation of a quintic hypersurface in CP^4:

$$X_1X_2X_3X_4X_5 f_\Delta\left(\frac{X_1}{X_5}, \frac{X_2}{X_5}, \frac{X_3}{X_5}, \frac{X_4}{X_5}\right) = \sum_{\nu_i \in M \cap \Delta} a_{\nu_i} (X_1)^{\nu_i^1+1} (X_2)^{\nu_i^2+1} (X_3)^{\nu_i^3+1} (X_4)^{\nu_i^4+1} (X_5)^{1-\sum_{j=1}^4 \nu_4^j}. \tag{4.32}$$

Coordinates $(\nu_i^1, \nu_i^2, \nu_i^3, \nu_i^4)$ of lattice points in Δ are given by solution of the following inequalities:

$$\nu_i^1 \geq -1, \;\; \nu_i^2 \geq -1, \;\; \nu_i^3 \geq -1, \;\; \nu_i^4 \geq -1, \;\; \nu_i^1 + \nu_i^2 + \nu_i^3 + \nu_i^4 \leq 1, \tag{4.33}$$

and the number of its solutions turns out to be 126. As can be seen from the form of (4.32), these lattice points generate all the degree 5 monomials in X_i. Therefore, (4.32) gives us the most general form of the defining equation of quintic hypersurface.

Next, we construct the mirror manifold (orbifold) of this quintic hypersurface by using toric geometry. The construction has an impressive feature. In short, it is done by exchanging the roles of M and N, or Δ and Δ^*. Hence the mirror manifold (orbifold) is given as the hypersurface of a complex 4-dimensional space obtained from the pair (Δ^*, Δ) (note here that the order is reversed). Let Y_i $(i = 1, 2, \ldots, 5)$ be homogeneous coordinates associated with the vertices u_i $(i = 1, 2, \ldots, 5)$ of Δ. Since Δ is a 5-gon in $\mathbf{R}^4$, the set $Z(\Delta)$, which is removed from $\mathbf{C}^5 = \{(Y_1, Y_2, \ldots, Y_5)\}$ in imposing the equivalence relation, turns out to be $\{\mathbf{0}\}$. Since u_i $(i = 1, 2, \ldots, 5)$ satisfy the unique linear relation:

$$u_1 + u_2 + u_3 + u_4 + u_5 = 0, \tag{4.34}$$

the equivalence relation imposed on $\mathbf{C}^5 - \{\mathbf{0}\}$ is given by

$$(Y_1, Y_2, Y_3, Y_4, Y_5) \sim (\lambda Y_1, \lambda Y_2, \lambda Y_3, \lambda Y_4, \lambda Y_5), \quad (\lambda \in \mathbf{C}^\times). \tag{4.35}$$

At first sight of these results, it may seem that the complex 4-dimensional space obtained from (Δ^*, Δ) also becomes CP^4. But we have to note here the fact that cones in the fan obtained from Δ are all singular. For example, the 4-dimensional cone $\sigma =< u_1, u_2, u_3, u_4 >_{\mathbf{R}_{\geq 0}}$ contains lattice points that cannot be represented as the integral linear combination of generators u_1, u_2, u_3, u_4. This can be observed from the fact that the determinant of the 4×4 matrix obtained from arranging u_i $(i = 1, 2, 3, 4)$ in vertical order exceeds 1:

$$\begin{vmatrix} 4 & -1 & -1 & -1 \\ -1 & 4 & -1 & -1 \\ -1 & -1 & 4 & -1 \\ -1 & -1 & -1 & 4 \end{vmatrix} = 125. \tag{4.36}$$

In such cases, we have to add information coming from singular characteristics. For this purpose, we introduce an abelian group $M_1 = \mathbf{Z}u_1 + \mathbf{Z}u_2 + \mathbf{Z}u_3 + \mathbf{Z}u_4 + \mathbf{Z}u_5$ generated by $u_1, u_2.u_3, u_4, u_5$. This M_1 is a subgroup of $M = \mathbf{Z}^4$ and quotient group $G = M/M_1$ acts on $CP^4 = \{(Y_1 : Y_2 : Y_3 : Y_4 : Y_5)\}$. G turns out to be isomorphic to $\left(\mathbf{Z}/(5\mathbf{Z})\right)^3$ as an abelian group. We use the local coordinate system of CP^4 to take a closer look at the action of G on CP^4. Let us recall the construction of local coordinates. We take a 4-dimensional cone σ in the fan obtained from Δ, determine generators of the dual 4-dimensional cone σ^* and associate local coordinates with the generators. For simplicity, we execute the process for the 4-dimensional cone $\sigma = < u_1, u_2, u_3, u_4 >_{\mathbf{R}_{\geq 0}}$. Generators of the dual cone σ^* in $N_{\mathbf{R}}$ are read off from column vectors of inverse matrix of the matrix in (4.36) and are given as follows.

$$w_1 = \left(\frac{2}{5}, \frac{1}{5}, \frac{1}{5}, \frac{1}{5}\right), \; w_2 = \left(\frac{1}{5}, \frac{2}{5}, \frac{1}{5}, \frac{1}{5}\right), \; w_3 = \left(\frac{1}{5}, \frac{1}{5}, \frac{2}{5}, \frac{1}{5}\right), \; w_4 = \left(\frac{1}{5}, \frac{1}{5}, \frac{1}{5}, \frac{2}{5}\right). \tag{4.37}$$

We obtain the local coordinate system (z_1, z_2, z_3, z_4) by associating z_i with the generator w_i. Since G is given by M/M_1, we can take e_j $(j = 1, \ldots, 4)$, the standard basis of M as generators of G. In short, the action of e^j on z_k is given by $z_k \to \exp(2\pi i\langle e_i, w_k\rangle)z_k$. Therefore, the action of e_j $(j = 1, \ldots, 4)$ on (z_1, z_2, z_3, z_4) is explicitly written down by using $\zeta = \exp(\frac{2\pi i}{5})$ as follows:

$$\begin{aligned}
&(z_1, z_2, z_3, z_4) \to (\zeta^2 z_1, \zeta z_2, \zeta z_3, \zeta z_4), \; (z_1, z_2, z_3, z_4) \to (\zeta z_1, \zeta^2 z_2, \zeta z_3, \zeta z_4),\\
&(z_1, z_2, z_3, z_4) \to (\zeta z_1, \zeta z_2, \zeta^2 z_3, \zeta z_4), \; (z_1, z_2, z_3, z_4) \to (\zeta z_1, \zeta z_2, \zeta z_3, \zeta^2 z_4).
\end{aligned} \tag{4.38}$$

Then we lift this action to action on homogeneous coordinates by using the relation $(z_1, z_2, z_3, z_4) = (\frac{Y_1}{Y_5}, \frac{Y_2}{Y_5}, \frac{Y_3}{Y_5}, \frac{Y_4}{Y_5})$. The result is given by

$$\begin{aligned}
&(Y_1 : Y_2 : Y_3 : Y_4 : Y_5) \to (\zeta^{m_1} Y_1 : \zeta^{m_2} Y_2 : \zeta^{m_3} Y_3 : \zeta^{m_4} Y_4 : \zeta^{m_5} Y_5),\\
&(\zeta = \exp\left(\frac{2\pi i}{5}\right), \; m_j \in \mathbf{Z}, \; \sum_{j=1}^{5} m_j \equiv 0 \mod 5).
\end{aligned} \tag{4.39}$$

In conclusion, the complex 4-dimensional space obtained from the pair (Δ^*, Δ) is the quotient space CP^4/G whose point is given by the equivalence class under the above action of G on CP^4. Since this space has orbifold singularities, it is a complex 4-dimensional orbifold.

Now, we determine defining equation of Calabi–Yau hypersurface in this complex 4-dimensional orbifold. We follow the recipe used in the case of a quintic hypersurface in CP^4. Since the roles of Δ and Δ^* are exchanged, it is given by

$$f_{\Delta^*}(t_1, t_2, \ldots, t_n) = \sum_{\nu_i \in N \cap \Delta^*} b_{\nu_i} t^{\nu_i} = 0. \tag{4.40}$$

Since the lattice points in Δ^* are given by $\mathbf{0}$ and ν_i $(i = 1, \ldots, 5)$, it is explicitly written as

$$f_{\Delta^*}(t_1, t_2, t_3, t_4) = b_1 t_1 + b_2 t_2 + b_3 t_3 + b_4 t_4 + b_5 \frac{1}{t_1 t_2 t_3 t_4} + b_6. \tag{4.41}$$

In order to see how t_i's are related to homogeneous coordinates, we use the local coordinate system (z_1, z_2, z_3, z_4) associated with generators w_1, w_2, w_3, w_4. t_i is associated with f_i, which is the standard base of $N_{\mathbf{R}}$, and f_i is represented as the following linear combination of w_i's:

$$f^1 = 4w_1 - w_2 - w_3 - w_4,\ f^2 = -w_1 + 4w_2 - w_3 - w_4,$$
$$f^3 = -w_1 - w_2 + 4w_3 - w_4,\ f^4 = -w_1 - w_2 - w_3 + 4w_4, \quad (4.42)$$

as can be read off from (4.37). Hence we obtain the relation $t_i = \frac{(z_i)^5}{z_1 z_2 z_3 z_4}$. On the other hand, local coordinates are related to homogeneous coordinates via the relation $z_i = \frac{Y_i}{Y_5}$, and t_i is represented by using homogeneous coordinates as follows:

$$t_i = \frac{(Y_i)^5}{Y_1 Y_2 Y_3 Y_4 Y_5}, \quad (i = 1, 2, 3, 4). \quad (4.43)$$

We plug (4.43) into (4.40) and multiply the result by $Y_1 Y_2 Y_3 Y_4 Y_5$ to obtain a homogeneous polynomial. In this way, we obtain the following defining equation of Calabi–Yau hypersurface in CP^4/G:

$$b_1(Y_1)^5 + b_2(Y_2)^5 + b_3(Y_3)^5 + b_4(Y_4)^5 + b_5(Y_5)^5 + b_6 Y_1 Y_2 Y_3 Y_4 Y_5 = 0. (4.44)$$

The orbifold CP^4/G and the above defining Eq. (4.44) are essentially the same as the construction of mirror manifolds (orbifolds) of a quintic hypersurface used in the paper of Candelas et al. This is the outline of construction of a pair of complex 3-dimensional Calabi–Yau manifolds (orbifolds) by using toric geometry.

4.2.2 Blow-Up (Resolution of Singularity) and Hodge Number

In this subsection, we proceed with our discussions, limiting our scope to the $n = 4$ case, i.e., $M = \mathbf{Z}^4$, $N = \mathbf{Z}^4$. We introduce here some new notations in order to summarize discussions in the previous subsection. We denote the 4-dimensional complex space obtained from the pair (Δ, Δ^*) of a reflexive convex polytope and its dual polytope by $P_{(\Delta,\Delta^*)}$. We also denote by $X_{(\Delta,\Delta^*)}$ the Calabi–Yau hypersurface in $P_{(\Delta,\Delta^*)}$, which was constructed in the previous subsection. Under these notations, our construction in the previous subsection is summarized as follows:

- $X_{(\Delta^*,\Delta)}$ is the mirror manifold (orbifold) of $X_{(\Delta,\Delta^*)}$.

As was discussed in Chap. 1, for a complex 3-dimensional Calabi–Yau manifold X and its mirror complex 3-dimensional Calabi–Yau manifold X^*, the symmetry relation of Hodge numbers,

$$h^{p,q}(X) = h^{3-p,q}(X^*), \quad (4.45)$$

must hold. Then does the relation $h^{p,q}(X_{(\Delta^*,\Delta)}) = h^{3-p,q}(X^{X_{(\Delta,\Delta^*)}})$ hold for the above $X_{(\Delta^*,\Delta)}$ and $X_{(\Delta,\Delta^*)}$? Unfortunately, it does not hold just as it is. This is because there exists the possibility that $X_{(\Delta^*,\Delta)}$ or $X_{(\Delta,\Delta^*)}$ may be the orbifold with singularities.

Therefore we apply resolution of singularity (blow-up), which is an technique in algebraic geometry, to $X_{(\Delta,\Delta^*)}$ and $X_{(\Delta^*,\Delta)}$, and obtain non-singular manifolds $\widetilde{X}_{(\Delta,\Delta^*)}$ and $\widetilde{X}_{(\Delta^*,\Delta)}$. Then the relation

$$h^{p,q}(\widetilde{X}_{(\Delta,\Delta^*)}) = h^{3-p,q}(\widetilde{X}_{(\Delta^*,\Delta)}) \tag{4.46}$$

indeed holds. In this subsection, we pick up a pair of reflexive polytopes that were discussed in the previous subsection and confirm the relation

$$h^{1,1}(\widetilde{X}_{(\Delta,\Delta^*)}) = h^{2,1}(\widetilde{X}_{(\Delta^*,\Delta)}),\ h^{2,1}(\widetilde{X}_{(\Delta,\Delta^*)}) = h^{1,1}(\widetilde{X}_{(\Delta^*,\Delta)}), \tag{4.47}$$

which was most non-trivial among (4.46) (the remaining relations follow from the fact that $\widetilde{X}_{(\Delta,\Delta^*)}$ and $\widetilde{X}_{(\Delta^*,\Delta)}$ are complex 3-dimensional Calabi–Yau manifolds).

We first discuss the pair (Δ, Δ^*) whose 4-dimensional complex space $P_{(\Delta,\Delta^*)}$ is given by CP^4. In this case, $P_{(\Delta,\Delta^*)} = CP^4$ is non-singular and a quintic hypersurface $X_{(\Delta,\Delta^*)}$ in CP^4 is generically non-singular. Hence $\widetilde{X}_{(\Delta,\Delta^*)}$ is given by $X_{(\Delta,\Delta^*)}$. It is well known from the Lefschetz hyperplane theorem that any element of $H^{1,1}(X_{(\Delta,\Delta^*)}, \mathbf{C})$ is induced from $H^{1,1}(P_{(\Delta,\Delta^*)}, \mathbf{C})$ via the pull-back i^* of inclusion map $i : X_{(\Delta,\Delta^*)} \hookrightarrow P_{(\Delta,\Delta^*)}$ and that i^* is injective. Therefore we have $h^{1,1}(\widetilde{X}_{(\Delta,\Delta^*)}) = h^{1,1}(X_{(\Delta,\Delta^*)}) = h^{1,1}(P_{(\Delta,\Delta^*)}) = h^{1,1}(CP^4)$. We already know that $h^{1,1}(CP^4)$ equals 1, which leads us to the conclusion that $h^{1,1}(\widetilde{X}_{(\Delta,\Delta^*)})$ also equals 1. At this stage, we mention the general result on $h^{1,1}(P_{(\Delta,\Delta^*)})$.[1] Let v_i $(i = 1, 2, \ldots, r)$, $(r > n)$ be vertices of Δ^*. Then we have the following exact sequence:

$$0 \to M \otimes_{\mathbf{Z}} \mathbf{C} \xrightarrow{\phi} \bigoplus_{j=1}^{r} \mathbf{C}v_j \to H^{1,1}(P_{(\Delta,\Delta^*)}, \mathbf{C}) \to 0. \tag{4.48}$$

Here $\oplus_{j=1}^{r}\mathbf{C}v_j$ is an abstract complex r-dimensional vector space whose bases are given by v_j's and ϕ is defined by $\phi(m) = \sum_{j=1}^{r}\langle m, v_j\rangle v_j$. Since $\dim_{\mathbf{C}}(M \otimes_{\mathbf{Z}} \mathbf{C}) = n$, we can see from the exact sequence that $h^{1,1}(P_{(\Delta,\Delta^*)}) = \dim_{\mathbf{C}}(H^{1,1}(P_{(\Delta,\Delta^*)}, \mathbf{C}))$ is given by $r - n$. When $P_{(\Delta,\Delta^*)}$ is CP^4, we also obtain $h^{1,1}(CP^4) = 5 - 4 = 1$ from this result.

Next, we discuss $h^{1,1}(\widetilde{X}_{(\Delta^*,\Delta)})$. As we have seen in the previous section, $X_{(\Delta^*,\Delta)}$ is the hypersurface in $P_{(\Delta^*,\Delta)}$. Since $P_{(\Delta^*,\Delta)}$ is an orbifold with singularities, $X_{(\Delta^*,\Delta)}$ is also singular. Therefore, $\widetilde{X}_{(\Delta^*,\Delta)}$ is different from $X_{(\Delta^*,\Delta)}$. In order to construct $\widetilde{X}_{(\Delta^*,\Delta)}$, we first construct $\widetilde{P}_{(\Delta^*,\Delta)}$ by resolving singularities of $P_{(\Delta^*,\Delta)}$. Then $X_{(\Delta^*,\Delta)}$ is given by a hypersurface in $\widetilde{P}_{(\Delta^*,\Delta)}$ whose defining equation is the same as that of $X_{(\Delta^*,\Delta)}$. In this case, the relation $h^{1,1}(\widetilde{X}_{(\Delta^*,\Delta)}) = h^{1,1}(\widetilde{P}_{(\Delta^*,\Delta)})$ also holds. Hence we have only to determine $h^{1,1}(\widetilde{P}_{(\Delta^*,\Delta)})$. Since Δ is a polytope with five vertices, the exact sequence (4.48) gives us $h^{1,1}(P_{(\Delta^*,\Delta)}) = 5 - 4 = 1$. What we have to do next is to classify sets of singularities in $P_{(\Delta^*,\Delta)}$, but we can execute this process by analyzing the distribution of lattice points in Δ. This recipe comes from toric geometry, and

[1]In the case when $P_{(\Delta,\Delta^*)}$ is singular, this $h^{1,1}(P_{(\Delta,\Delta^*)})$ should be interpreted as the rank of the Chow ring $CH^1(P_{(\Delta,\Delta^*)})$, but in this book, we avoid using this technical term in algebraic geometry.

we explain briefly the outline. Let $\Delta_{(u_{i_0},u_{i_1},\ldots,u_{i_l})}$ be l-dimensional face of Δ, which is a convex hull of $l+1$ vertices u_{i_j} ($j = 0, 1, 2, \ldots, l$ and $i_0 < i_1 < \ldots < i_l$) of Δ. This $\Delta_{(u_{i_0},u_{i_1},\ldots,u_{i_l})}$ corresponds to the following complex $3-l$-dimensional subset of $P_{(\Delta^*,\Delta)}$:

$$T_{(u_{i_0},u_{i_1},\ldots,u_{i_l})} = \{(Y_0 : Y_1 : Y_2 : Y_3 : Y_4 : Y_5) \mid Y_j = 0 \ (j = i_m),\ Y_j \neq 0 \ (j \neq i_m)\ \}/G. \quad (4.49)$$

The general theory of toric geometry tells us that $T_{(u_{i_0},u_{i_1},\ldots,u_{i_l})}$ becomes a set of singularities when $\Delta_{(u_{i_0},u_{i_1},\ldots,u_{i_l})}$ contains lattice points in its interior. Resolution of singularities of $T_{(u_{i_0},u_{i_1},\ldots,u_{i_l})}$ is done by adding these interior lattice points as vertices of $\Delta_{(u_{i_0},u_{i_1},\ldots,u_{i_l})}$. We have to add some more explanation on adding vertices. Adding internal lattice points of $l(\leq 3)$-dimensional faces of Δ as vertices does not change the shape of polytope Δ, but it induces new homogeneous coordinates of $P_{(\Delta^*,\Delta)}$ in associating homogeneous coordinates with vertices of Δ. Of course, even if one new homogneous coordinate is added, we have one more linear relations among vertices. Therefore, the dimension of the resulting complex space remains four. It is known in toric geometry that this operation resolves orbifold singularities of $T_{(u_{i_0},u_{i_1},\ldots,u_{i_l})}$. Let us apply this recipe to $P_{(\Delta^*,\Delta)}$ in consideration. What we have to do is to add lattice points in Δ except for $\mathbf{0}$ and $u_1, u_2, \ldots, u_5$ as new vertices. The number of such lattice points is given by $126 - 1 - 5 = 120$, as was discussed in the previous subsection. By applying the exact sequence (4.48), $h^{1,1}$ of the space obtained after resolving all the singularities turns out to be $120 + 5 - 4 = 121$. But there are some subtleties. In constructing $\widetilde{X}_{(\Delta^*,\Delta)}$, we don't have to resolve the 0-dimensional set of singularities $T_{(u_{i_0},u_{i_1},u_{i_2},u_{i_3})}$. Since $\widetilde{X}_{(\Delta^*,\Delta)}$ is a complex codimension 1 hypersurface, we can avoid the 0-dimensional set of singularities by adequately choosing the defining equation of the hypersurface. In contrast, we have to resolve the set of singularities whose dimension is more than 1 because it always has a non-trivial intersection with $\widetilde{X}_{(\Delta^*,\Delta)}$. Since the five 3-dimensional faces $\Delta_{(u_1,u_2,u_3,u_4)}$, $\Delta_{(u_1,u_2,u_3,u_5)}$, $\Delta_{(u_1,u_2,u_4,u_5)}$, $\Delta_{(u_1,u_3,u_4,u_5)}$, $\Delta_{(u_2,u_3,u_4,u_5)}$ have four internal lattice points respectively (we leave confirmation to the readers as an exercise), the number of added vertices becomes $120 - 4 \times 5 = 100$. Then we obtain

$$h^{1,1}(\widetilde{X}_{(\Delta^*,\Delta)}) = h^{1,1}(\widetilde{P}_{(\Delta^*,\Delta)}) = 1 + 100 = 101. \quad (4.50)$$

Let us summarize the discussions so far by introducing new notations. Let $l(\Delta)$ be the number of lattice points in the 4-dimensional reflexive polytope Δ. We denote each 3-dimensional face of Δ by Γ and represent the number of internal lattice points of Γ as $l^*(\Gamma)$. In the case of the quintic hypersurface $\widetilde{X}_{(\Delta,\Delta^*)}$ in CP^4, $h^{1,1}(\widetilde{X}_{(\Delta,\Delta^*)})$ was given by $5 - 4 = 1$, but we reinterpret this 1 as $6 - 5$ by adding 1 to 5 and 4 respectively. The number 6 is the number of lattice points of Δ^* since Δ^* has only $\mathbf{0}$ and v_i's as lattice points. Moreover, $l^*(\Gamma) = 0$ holds for each 3-dimensional face of Δ^*. Therefore, we can rewrite the derivation of $h^{1,1}(\widetilde{X}_{(\Delta,\Delta^*)})$ as follows:

$$h^{1,1}(\widetilde{X}_{(\Delta,\Delta^*)}) = l(\Delta^*) - 5 - \sum_{\Gamma\subset\Delta^*} l^*(\Gamma). \tag{4.51}$$

In the case of $\widetilde{X}_{(\Delta^*,\Delta)}$, $h^{1,1}(\widetilde{X}_{(\Delta^*,\Delta)})$ is obtained by subtracting 4 from the number of vertices of Δ and adding $l(\Delta) - ((\text{number of vertices of } \Delta) + 1) - \sum_{\Gamma\subset\Delta} l^*(\Gamma)$ to the result. Hence we obtain the following formula:

$$h^{1,1}(\widetilde{X}_{(\Delta^*,\Delta)}) = l(\Delta) - 5 - \sum_{\Gamma\subset\Delta} l^*(\Gamma). \tag{4.52}$$

If we compare (4.51) with (4.52), we can easily see that these are symmetric under exchange of Δ and Δ^*.

Let us turn to discussions on $h^{2,1}$. As was mentioned in Chap. 1, $h^{2,1}(X)$ of a 3-dimensional Calabi–Yau manifold X equals the number of degrees of freedom of deformation of the complex structure of X, or more intuitively, the number of substantial parameters in the defining equation of X. In the case of a quintic hypersurface in CP^4, the number of parameters in the defining equation is 126, which was computed in the previous subsection. Since multiplication by a non-zero constant does not change the hypersurface, we have to subtract 1 from 126. Moreover, we have to subtract the number of degrees of freedom of $\mathrm{Aut}(CP^4)$, the automorphism group of CP^4. $\mathrm{Aut}(CP^4)$ is the set of biholomorphic maps from CP^4 to CP^4, which are explicitly given by $(X_i \rightarrow \sum_{j=1}^5 a_i^j X_j)$, where $A = (a_i^j)$ is a 5×5 complex invertible matrix. Let $f(X_1 : \cdots : X_5)$ be the defining equation of a quintic hypersuface. Then $f(\sum_{j=1}^5 a_1^j X_j : \cdots : \sum_{j=1}^5 a_5^j X_j) = \tilde{f}(X_1 : \cdots : X_5)$ gives us another defining equation of a quintic hypersurface. But the map $(X_i \rightarrow \sum_{j=1}^5 a_i^j X_j)$ is a biholomorphic map between the quintic hypersurface defined by f and the one defined by $\tilde{f}$, and the complex structures of these two hypersurfaces are the same. This is the reason for the subtraction of $\dim_{\mathbf{C}}(\mathrm{Aut}(CP^4))$. Since $\dim_{\mathbf{C}}(\mathrm{Aut}(CP^4))$ is given by $5 \times 5 - 1 = 24$ (we have to subtract 1 by projective equivalence), $h^{2,1}$ of the quintic hypersurface turns out to be $126 - 1 - 24 = 101$. In this way, we can confirm $h^{2,1}(\widetilde{X}_{(\Delta,\Delta^*)}) = h^{1,1}(\widetilde{X}_{(\Delta^*,\Delta)})$ when $\widetilde{X}_{(\Delta^*,\Delta)}$ is a quintic hypersurface. In order to discuss more general cases, we introduce a formula of toric geometry which gives us the complex dimension of $\mathrm{Aut}(P_{(\Delta,\Delta^*)})$:

$$\dim_{\mathbf{C}}(\mathrm{Aut}(P_{(\Delta,\Delta^*)})) = 4 + \sum_{\Gamma\subset\Delta} l^*(\Gamma). \tag{4.53}$$

We note here that the resolution of singularities does not change the dimension of the automorphism group. We don't go into the proof of this formula, but in the case of $\mathrm{Aut}(CP^4)$, the 4 in the r.h.s. corresponds to the number of diagonal elements of $A = (a_i^j)$ up to overall multiplication of a non-zero constant, the four interior lattice points of $\Delta_{(u_1,u_2,u_3,u_4)}$ correspond to a_{5j} $(j = 1, 2, 3, 4)$, the four interior lattice points of $\Delta_{(u_1,u_2,u_3,u_5)}$ correspond to a_{4j} $(j = 1, 2, 3, 4)$, and so on. By using the formula, the previous computation $h^{2,1}(X_{(\Delta,\Delta^*)}) = 125 - 24 = 101$ is rewritten as follows:

$$h^{2,1}(\widetilde{X}_{(\Delta,\Delta^*)}) = l(\Delta) - 5 - \sum_{\Gamma \subset \Delta} l^*(\Gamma). \tag{4.54}$$

On the other hand, the defining equation of $\widetilde{X}_{(\Delta^*,\Delta)}$ does not change after the resolution of singularities, and we can apply the same recipe of counting $h^{2,1}$ to $\widetilde{X}_{(\Delta^*,\Delta)}$. As we have seen in the previous subsection, the number of parameters of the defining equation up to multiplication by a non-zero constant is given by 5. Since the 3-dimensional faces of Δ^* have no interior lattice points, (4.53) tells us that $\dim_{\mathbf{C}}(\mathrm{Aut}(P_{(\Delta^*,\Delta)}))$ equals 4. Hence we obtain $h^{2,1}(\widetilde{X}_{(\Delta^*,\Delta)}) = 5 - 4 = 1$, but we can obtain the same result by applying (4.54) with Δ and Δ^* interchanged:

$$h^{2,1}(\widetilde{X}_{(\Delta^*,\Delta)}) = l(\Delta^*) - 5 - \sum_{\Gamma \subset \Delta^*} l^*(\Gamma) = 1. \tag{4.55}$$

Thus we have confirmed the equality $h^{2,1}(\widetilde{X}_{(\Delta^*,\Delta)}) = h^{1,1}(\widetilde{X}_{(\Delta,\Delta^*)})$ when $\widetilde{X}_{(\Delta^*,\Delta)}$ is a quintic hypersurface. If we look back at the formulas that represent $h^{1,1}$ and $h^{2,1}$ in terms of toric geometry, we can see that the mirror symmetry of Hodge numbers between a quintic hypersurface $\widetilde{X}_{(\Delta,\Delta^*)}$ in CP^4 and its mirror manifold $\widetilde{X}_{(\Delta^*,\Delta)}$ is quite natural as if it is trivial.

In the remaining part of this subsection, we mention the well-known result on Hodge numbers of a complex 3-dimensional Calabi–Yau manifold $\widetilde{X}_{(\Delta,\Delta^*)}$ obtained from general pair (Δ, Δ^*) of reflexive convex polytopes. In the celebrated work [1] of 1994, Batyrev proved the following formulas:

$$\begin{aligned} h^{1,1}(\widetilde{X}_{(\Delta,\Delta^*)}) &= l(\Delta^*) - 5 - \sum_{\Gamma^* \subset \Delta^*} l^*(\Gamma^*) + \sum_{\Theta^* \subset \Delta^*} l^*(\Theta^*) l^*(\widehat{\Theta}^*), \\ h^{2,1}(\widetilde{X}_{(\Delta,\Delta^*)}) &= l(\Delta) - 5 - \sum_{\Gamma \subset \Delta} l^*(\Gamma) + \sum_{\Theta \subset \Delta} l^*(\Theta) l^*(\widehat{\Theta}). \end{aligned} \tag{4.56}$$

In the above formulas, Γ (Γ^*) is the 3-dimensional face of Δ (Δ^*), Θ (Θ^*) is the 2-dimensional face of Δ (Δ^*) and $\widehat{\Theta}$ ($\widehat{\Theta}^*$) is the 1-dimensional face of Δ^* (Δ), which is dual to Θ (Θ^*) (for example, $\widehat{\Theta}$ is obtained as the convex hull of two vertices of Δ^*, each of which is the defining equation of the 2-dimensional plane in $M_{\mathbf{R}}$ that contains Θ). As has already been used, l represents the operation of counting lattice points in a polytope and l^* represents the operation of counting interior lattice points of a face. In the example of this subsection, the third term in the r.h.s. vanishes. The mirror symmetry of Hodge numbers,

$$h^{1,1}(\widetilde{X}_{(\Delta,\Delta^*)}) = h^{2,1}(\widetilde{X}_{(\Delta^*,\Delta)}), \; h^{2,1}(\widetilde{X}_{(\Delta,\Delta^*)}) = h^{1,1}(\widetilde{X}_{(\Delta^*,\Delta)}), \tag{4.57}$$

is obvious since the r.h.s. of the second line of (4.56) is obtained from the r.h.s. of the first line by changing Δ^* into Δ. In this way, toric geometry systematically produces mirror pairs of complex 3-dimensional Calabi–Yau manifolds.

4.3 Details of B-Model Computation

In Chap. 1, we outlined the computation of Kähler deformation Yukawa coupling of a quintic hypersurface in CP^4 by using mirror symmetry (we call this process B-model computation). In this section, we give a detailed exposition of the B-model computation in the case of a quintic hypersurface with the aid of toric geometry.

4.3.1 *Derivation of Differential Equations Satisfied by Period Integrals*

We already know from discussions on the topological sigma model (B-model) that Yukawa coupling associated with complex structure deformation of a complex 3-dimensional Calabi–Yau manifold X is given by the following formula:

$$\langle \mathcal{O}_{u_{(\alpha)}} \mathcal{O}_{u_{(\beta)}} \mathcal{O}_{u_{(\gamma)}} \rangle = \int_X \Omega \wedge \left(u_{(\alpha)} \wedge u_{(\beta)} \wedge u_{(\gamma)} \wedge \Omega \right). \tag{4.58}$$

Here, Ω is the holomorphic (3, 0)-form of X, and $u_{(\alpha)}$, $u_{(\beta)}$, $u_{(\gamma)}$ are elements of $H^1(X, T'X)$. We have explained that $H^1(X, T'X) \simeq H^{2,1}(X, \mathbf{C})$ is identified with degrees of freedom of substantial deformation of the defining equation of X. Ω depends on parameters that describe this deformation. Let a be the deformation parameter of the defining equation that correspond to $u \in H^1(X, T'X)$. Then we can use the Kodaira–Spencer equation, which was mentioned in Chap. 1:

$$u \wedge \Omega = \frac{\partial \Omega}{\partial a} + f \cdot \Omega, \tag{4.59}$$

where f is some holomorphic function of deformation parameters. By combining this equation with (4.58), Yukawa coupling is represented by

$$\langle \mathcal{O}_{u_{(\alpha)}} \mathcal{O}_{u_{(\beta)}} \mathcal{O}_{u_{(\gamma)}} \rangle = \int_X \Omega \wedge \frac{\partial^3 \Omega}{\partial a_{(\alpha)} \partial a_{(\beta)} \partial a_{(\gamma)}}, \tag{4.60}$$

which already appeared in Chap. 1. Here, $a_{(*)}$ is the deformation parameter that corresponds to $u_{(*)}$. This is the formula of B-model Yukawa coupling that we use in this section. In order to proceed further, we take basis α_i $(i = 1, 2, \ldots, h;\ h = \dim_{\mathbf{C}}(H^3(X, \mathbf{C}))$ of $H^3(X, \mathbf{C})$ and expand Ω in terms of these bases:

$$\Omega = \sum_{i=1}^{h} \left(\int_{A_i} \Omega\right) \alpha_i, \tag{4.61}$$

where A_i is a real 3-dimensional homology cycle of X and satisfies $\int_{A_i} \alpha_j = \delta_{ij}$. The expansion coefficient $\int_{A_i} \Omega$ is called the period integral of X and is a function of deformation parameters. It plays a central role in the B-model computation of Yukawa coupling, as was mentioned in Chap. 1. In this section, we first derive the differential equation satisfied by the period integrals of the Calabi–Yau manifold $\widetilde{X}_{(\Delta^*,\Delta)}$ when $\widetilde{X}_{(\Delta,\Delta^*)}$ is a quintic hypersurface in CP^4. Of course, $\widetilde{X}_{(\Delta^*,\Delta)}$ is the mirror Calabi–Yau manifold of the quintic hypersurface. We start by defining the period integral of $\widetilde{X}_{(\Delta^*,\Delta)}$ as the integral of the meromorphic $(4,0)$-form of $P_{(\Delta^*,\Delta)}$. Let us look back at the construction process of the Calabi–Yau manifold $X_{(\Delta^*,\Delta)}$ in $P_{(\Delta^*,\Delta)}$. We first introduced $\mathbf{C}^\times$-valued coordinates t_j that are associated with the standard basis f_j of N and constructed the defining equation of the Calabi–Yau hypersurface in the following form:

$$f_{\Delta^*}(t_*; b_*) = \sum_{v_i \in N \cap \Delta^*} b_{v_i} t^{v_i} = b_1 t_1 + b_2 t_2 + b_3 t_3 + b_4 t_4 + b_5 \frac{1}{t_1 t_2 t_3 t_4} + b_6 \qquad (4.62)$$

Since $(\mathbf{C}^\times)^4 = \{((t_1, t_2, t_3, t_4) \mid t_i \in \mathbf{C}^\times\}$ is a dense subset of $P_{(\Delta^*,\Delta)}$, the meromorphic $(4,0)$-form of $P_{(\Delta^*,\Delta)}$ can be given as the meromorphic $(4,0)$-form of $(\mathbf{C}^\times)^4$. With these considerations, we introduce the following definition of the period integral of $\widetilde{X}_{(\Delta^*,\Delta)}$, which is standard in toric geometry:

$$\Pi_i(b_*) = \int_{\gamma_i} \frac{b_6}{b_1 t_1 + b_2 t_2 + b_3 t_3 + b_4 t_4 + b_5 \frac{1}{t_1 t_2 t_3 t_4} + b_6} \prod_{j=1}^{4} \frac{dt_j}{t_j}] = \int_{\gamma_i} \frac{b_6}{f_{\Delta^*}(t_*; b_*)} \prod_{j=1}^{4} \frac{dt_j}{t_j}, \qquad (4.63)$$

where γ_i is a real 4-dimensional homology cycle that wraps around the Calabi–Yau hypersurface defined by $f_{\Delta^*}(t_*; b_*) = 0$. By choosing the cycle γ_i adequately, we can make $\Pi_i(b_*)$ coincide with $\int_{A_i} \Omega$. With this set-up, we derive the differential equation satisfied by $\Pi_i(b_*)$. We first show that $\Pi_i(b_*)$ satisfies the following equations:

$$\begin{aligned} &\left(\sum_{j=1}^{6} b_i \frac{\partial}{\partial b_i}\right) \Pi_i(b_*) = 0, \\ &\left(b_i \frac{\partial}{\partial b_i} - b_5 \frac{\partial}{\partial b_5}\right) \Pi_i(b_*) = 0, \;\; (i = 1,2,3,4), \\ &\left(\frac{\partial}{\partial b_1} \frac{\partial}{\partial b_2} \frac{\partial}{\partial b_3} \frac{\partial}{\partial b_4} \frac{\partial}{\partial b_5} - \left(\frac{\partial}{\partial b_6}\right)^5\right) \left(\frac{\Pi_i(b_*)}{b_6}\right) = 0. \end{aligned} \qquad (4.64)$$

Fundamentally, these equations are proved by using the fact that the order of integration of the parametrized function and the partial differential of parameters can be exchangeable. The first equation is proved as follows:

$$\begin{aligned} &\left(\sum_{j=1}^{6} b_i \frac{\partial}{\partial b_i}\right) \Pi_i(b_*) \\ &= \left(\sum_{j=1}^{6} b_i \frac{\partial}{\partial b_i}\right) \int_{\gamma_i} \frac{b_6}{f_{\Delta^*}(t_*; b_*)} \prod_{j=1}^{4} \frac{dt_j}{t_j} \end{aligned}$$

$$= \int_{\gamma_i} \left(b_6 \left(\sum_{j=1}^{6} b_i \frac{\partial}{\partial b_i} \right) \frac{1}{f_{\Delta^*}(t_*; b_*)} + \frac{1}{f_{\Delta^*}(t_*; b_*)} \left(\sum_{j=1}^{6} b_i \frac{\partial}{\partial b_i} \right) b_6 \right) \prod_{j=1}^{4} \frac{dt_j}{t_j}$$

$$= \int_{\gamma_i} \left(-\frac{f_{\Delta^*}(t_*; b_*)}{(f_{\Delta^*}(t_*; b_*))^2} + \frac{1}{f_{\Delta^*}(t_*; b_*)} \right) b_6 \prod_{j=1}^{4} \frac{dt_j}{t_j}$$

$$= 0. \tag{4.65}$$

In the fourth line, we used the relation $\frac{\partial}{\partial b_i} \frac{1}{f_{\Delta^*}(t_*; b_*)} = -\frac{t^{\nu_i}}{(f_{\Delta^*}(t_*; b_*))^2}$. We then turn to the second equation. We prove it by using the fact that the residue integral is invariant under the change of integration variables. We fix $i \in \{1, 2, 3, 4\}$ and consider change of variable $t_i \to e^{\varepsilon} t_i$. For example, we set $i = 1$. Since this operation does not change $\Pi(b_*)$, we obtain the following relation:

$$\Pi_i(b_*) = \int_{\gamma_i} \frac{b_6}{b_1 t_1 + b_2 t_2 + b_3 t_3 + b_4 t_4 + b_5 \frac{1}{t_1 t_2 t_3 t_4} + b_6} \prod_{j=1}^{4} \frac{dt_j}{t_j}$$

$$= \int_{\gamma_i} \frac{b_6}{e^{\varepsilon} b_1 t_1 + b_2 t_2 + b_3 t_3 + b_4 t_4 + e^{-\varepsilon} b_5 \frac{1}{t_1 t_2 t_3 t_4} + b_6} \prod_{j=1}^{4} \frac{dt_j}{t_j}. \tag{4.66}$$

By using this relation, we can prove the second equation as follows:

$$\frac{\partial}{\partial \varepsilon} \int_{\gamma_i} \frac{b_6}{e^{\varepsilon} b_1 t_1 + b_2 t_2 + b_3 t_3 + b_4 t_4 + e^{-\varepsilon} b_5 \frac{1}{t_1 t_2 t_3 t_4} + b_6} \prod_{j=1}^{4} \frac{dt_j}{t_j} |_{\varepsilon=0} = 0,$$

$$\Longleftrightarrow \int_{\gamma_i} \left(\frac{-b_1 t_1 + b_5 \frac{1}{t_1 t_2 t_3 t_4}}{(f_{\Delta^*}(t_*; b_*))^2} \right) b_6 \prod_{j=1}^{4} \frac{dt_j}{t_j} = 0,$$

$$\Longleftrightarrow \int_{\gamma_i} \left(\left(b_1 \frac{\partial}{\partial b_1} - b_5 \frac{\partial}{\partial b_5} \right) \frac{b_6}{(f_{\Delta^*}(t_*; b_*))} \right) \prod_{j=1}^{4} \frac{dt_j}{t_j} = 0,$$

$$\Longleftrightarrow \left(b_1 \frac{\partial}{\partial b_1} - b_5 \frac{\partial}{\partial b_5} \right) \int_{\gamma_i} \left(\frac{b_6}{(f_{\Delta^*}(t_*; b_*))} \right) \prod_{j=1}^{4} \frac{dt_j}{t_j} = 0,$$

$$\Longleftrightarrow \left(b_1 \frac{\partial}{\partial b_1} - b_5 \frac{\partial}{\partial b_5} \right) \Pi_i(b_*) = 0. \tag{4.67}$$

Proof of the equation for the other i goes in the same way. We can prove the third equation in (4.64) by the following direct computation:

$$\left(\frac{\partial}{\partial b_1} \frac{\partial}{\partial b_2} \frac{\partial}{\partial b_3} \frac{\partial}{\partial b_4} \frac{\partial}{\partial b_5} - \left(\frac{\partial}{\partial b_6} \right)^5 \right) \left(\frac{\Pi_i(b_*)}{b_6} \right)$$

$$\int_{\gamma_i}\left(\frac{\partial}{\partial b_1}\frac{\partial}{\partial b_2}\frac{\partial}{\partial b_3}\frac{\partial}{\partial b_4}\frac{\partial}{\partial b_5}-\left(\frac{\partial}{\partial b_6}\right)^5\right)\frac{1}{(f_{\Delta^*}(t_*;b_*))}\prod_{j=1}^{4}\frac{dt_j}{t_j}$$
$$=(-1)^5\cdot 5!\int_{\gamma_i}\left(\frac{t_1t_2t_3t_4\cdot\frac{1}{t_1t_2t_3t_4}-1}{(f_{\Delta^*}(t_*;b_*))^6}\right)\prod_{j=1}^{4}\frac{dt_j}{t_j}=0. \tag{4.68}$$

Now, we determine the explicit form of $\Pi_i(b_*)$ as a function by using the first two equations of (4.64). We first expand $\Pi_i(b_*)$ as power series in b_i's:

$$\Pi_i(b_*)=\sum_{n_j\in\mathbf{Z}}P^i_{n_1n_2n_3n_4n_5n_6}(b_1)^{n_1}(b_2)^{n_2}(b_3)^{n_3}(b_4)^{n_4}(b_5)^{n_5}(b_6)^{n_6}. \tag{4.69}$$

Then the first two equations reduce to the following conditions:

$$P^i_{n_1n_2n_3n_4n_5n_6}\neq 0\Longrightarrow n_1+n_2+n_3+n_4+n_5+n_6=0$$
$$\text{and } n_1=n_2=n_3=n_4=n_5. \tag{4.70}$$

From these condtions, we can conclude that Π_i depends only on the variable $z=\frac{b_1b_2b_3b_4b_5}{(b_6)^5}$. Therefore we rewrite the third equation of (4.64),

$$\Big((b_1\partial_1)(b_2\partial_2)(b_3\partial_3)(b_4\partial_4)(b_5\partial_5)-(b_1b_2b_3b_4b_5)(\partial_6)^5\Big)\left(\frac{1}{b_6}\Pi_i\right)=0, \tag{4.71}$$

into the differential equation satisfied by $\Pi_i(z)$ (from now on, we abbreviate $\frac{\partial}{\partial b_i}$ as ∂_i). By using the relation $\partial_6(\frac{1}{(b_6)^i}f(z))=(-i)\frac{1}{(b_6)^{i+1}}f(z)+\frac{1}{(b_6)^i}\partial_6 f(z)=\frac{1}{(b_6)^{i+1}}(b_6\partial_6-i)f(z)$, we obtain

$$(\partial_6)^5\left(\frac{1}{b_6}\Pi_i(z)\right)=\frac{1}{(b_6)^6}(b_6\partial_6-5)(b_6\partial_6-4)(b_6\partial_6-3)(b_6\partial_6-2)(b_6\partial_6-1)\Pi_i(z)$$
$$=\frac{1}{(b_6)^6}\left(\prod_{j=1}^{5}(b_6\partial_6-j)\right)\Pi_i(z). \tag{4.72}$$

Then we can rewrite Eq. (4.71) into the form:

$$\frac{1}{b_6}\left((b_1\partial_1)(b_2\partial_2)(b_3\partial_3)(b_4\partial_4)(b_5\partial_5)-z\prod_{j=1}^{5}(b_6\partial_6-j)\right)\Pi_i(z)=0. \tag{4.73}$$

If we take note of the relations:

$$b_i\partial_i f(z)=z\frac{df(z)}{dz}\ (i=1,2,3,4,5),\quad b_6\partial_6 f(z)=-5z\frac{df(z)}{dz}, \tag{4.74}$$

we can further rewrite the equation into the following final form:

$$\frac{1}{b_6}\left(\left(z\frac{d}{dz}\right)^5 - z\prod_{j=1}^{5}\left(-5z\frac{d}{dz}-j\right)\right)\Pi_i(z) = 0$$

$$\Longleftrightarrow \frac{z\frac{d}{dz}}{b_6}\left(\left(z\frac{d}{dz}\right)^4 + 5z\prod_{j=1}^{4}\left(5z\frac{d}{dz}+j\right)\right)\Pi_i(z) = 0. \quad (4.75)$$

If we ignore $\frac{z\frac{d}{dz}}{b_6}$ at the left end and change the variable z into $w = -5^5 z$, it becomes,

$$\left(\left(w\frac{d}{dw}\right)^4 - w\prod_{j=1}^{4}\left(w\frac{d}{dw}+\frac{j}{5}\right)\right)\Pi_i(w) = 0, \quad (4.76)$$

and coincides with the differential equation that appeared in Chap. 1. This is the derivation of the differential equation satisfied by period integral of $\widetilde{X}_{(\Delta^*,\Delta)}$ when $\widetilde{X}_{(\Delta,\Delta^*)}$ is a quintic hypersurface in CP^4. As some readers might have already noticed, this derivation uses only the information on $X_{(\Delta^*,\Delta)}$ instead of $\widetilde{X}_{(\Delta^*,\Delta)}$. In other words, we don't need information that comes from the resolution of singularities. This is because the B-model only depends on deformation of the complex structure and the process of resolution of singularities only affects deformation of the Kähler structure.

4.3.2 Derivation of B-Model Yukawa Coupling

In this subsection, we determine the B-model Yukawa coupling explicitly when $\widetilde{X}_{(\Delta,\Delta^*)}$, the mirror manifold of $\widetilde{X}_{(\Delta^*,\Delta)}$, is a quintic hypersurface in CP^4. In this case, the formula (4.60) reduces to

$$Y_{uuu} = \int_{\widetilde{X}_{(\hat{\Delta}^*,\Delta)}} \Omega \wedge \frac{\partial^3}{\partial u^3}\Omega, \quad (4.77)$$

where u is the deformation parameter in the differential equation

$$\left(\left(\frac{d}{du}\right)^4 - 5e^u\prod_{j=1}^{4}\left(5\frac{d}{du}+j\right)\right)\Pi_i(u) = 0$$

$$\Longleftrightarrow \left((1-5^5e^u)\left(\frac{d}{du}\right)^4 - e^u\left(2\cdot 5^5\left(\frac{d}{du}\right)^3 + 7\cdot 5^4\left(\frac{d}{du}\right)^2\right.\right.$$

$$\left.\left. + 2\cdot 5^4\frac{d}{du} + 24\cdot 5\right)\right)\Pi_i(u) = 0. \quad (4.78)$$

Here, we remove $\frac{z\frac{d}{dz}}{b_6}$ from the left end of Eq. (4.75) and change the variable from z into $u = -\log(z)$.

Period integrals are related to the holomorphic (3, 0) form Ω of $\widetilde{X}_{(\Delta^*,\Delta)}$ via the following relation:

$$\Omega = \sum_{i=1}^{h} \Pi_i(u)\alpha_i, \tag{4.79}$$

where α_i $(i = 1, 2, \ldots, h(= \dim_{\mathbf{C}}(H^3(\widetilde{X}_{(\Delta^*,\Delta)}, \mathbf{C}))))$ are bases of the chomology group $H^3(\widetilde{X}_{(\Delta^*,\Delta)}, \mathbf{C})$. Since α_i depends only on the topology of $\widetilde{X}_{(\Delta^*,\Delta)}$ and not on deformation of the complex structure, (4.78) and (4.79) lead us to the following relation:

$$\left((1-5^5e^u)\left(\frac{\partial}{\partial u}\right)^4 - e^u\left(2\cdot 5^5\left(\frac{\partial}{\partial u}\right)^3 + 7\cdot 5^4\left(\frac{\partial}{\partial u}\right)^2 + 2\cdot 5^4\frac{\partial}{\partial u} + 24\cdot 5\right)\right)\Omega = 0. \tag{4.80}$$

By operating $\int_{\widetilde{X}_{(\Delta^*,\Delta)}} \Omega\wedge$ from the left on both sides of (4.80), we obtain the equation:

$$(1-5^5e^u)\int_{\widetilde{X}_{(\Delta^*,\Delta)}} \Omega \wedge \frac{\partial^4}{\partial u^4}\Omega = 2\cdot 5^5 e^u \int_{\widetilde{X}_{(\Delta^*,\Delta)}} \Omega \wedge \frac{\partial^3}{\partial u^3}\Omega, \tag{4.81}$$

where we used the relation $\int_{\widetilde{X}_{(\Delta^*,\Delta)}} \Omega \wedge \frac{\partial^j}{\partial u^j}\Omega = 0$ $(j = 0, 1, 2)$ (this follows from the fact that $\frac{d^j}{du^j}\Omega$ is a 3-form whose anti-holomorphic degree is at most two).

We can derive another equation,

$$\int_{\widetilde{X}_{(\Delta^*,\Delta)}} \Omega \wedge \frac{\partial^4}{\partial u^4}\Omega = 2\frac{d}{du}Y_{uuu}. \tag{4.82}$$

This is proved as follows.

By differentiating both sides of $\int_{\widetilde{X}_{(\Delta^*,\Delta)}} \Omega \wedge \frac{d^2}{du^2}\Omega = 0$ two times by u, we obtain

$$2\int_{\widetilde{X}_{(\Delta^*,\Delta)}} \frac{d}{du}\Omega \wedge \frac{\partial^3}{\partial u^3}\Omega + \int_{\widetilde{X}_{(\Delta^*,\Delta)}} \Omega \wedge \frac{d^4}{du^4}\Omega = 0, \tag{4.83}$$

where we used the equality $\int_{\widetilde{X}_{(\Delta^*,\Delta)}} \frac{\partial^2}{\partial u^2}\Omega \wedge \frac{\partial^2}{\partial u^2}\Omega = 0$. On the other hand, differentiating both sides of (4.77) by u results in

$$\frac{d}{du}Y_{uuu} = \int_{\widetilde{X}_{(\Delta^*,\Delta)}} \frac{\partial}{\partial u}\Omega \wedge \frac{\partial^3}{\partial u^3}\Omega + \int_{\widetilde{X}_{(\Delta^*,\Delta)}} \Omega \wedge \frac{\partial^4}{\partial u^4}\Omega. \tag{4.84}$$

By substituting $\int_{\widetilde{X}_{(\Delta^*,\Delta)}} \frac{\partial}{\partial u}\Omega \wedge \frac{\partial^3}{\partial u^3}\Omega = -\frac{1}{2}\int_{\widetilde{X}_{(\Delta^*,\Delta)}} \Omega \wedge \frac{\partial^4}{\partial u^4}\Omega$ into the above equation, we obtain (4.82).

Combination of (4.81) and (4.82) leads us to the following differential equation of the B-model Yukawa coupling Y_{uuu}:

$$(1-5^5e^u)\frac{d}{du}Y_{uuu} = 5^5e^uY_{uuu}. \tag{4.85}$$

Since (4.85) is a separable differential equation, it is easy to integrate it out. We finally obtain the B-model Yukawa coupling in the following form,

$$Y_{uuu} = \frac{C_0}{1-5^5e^u}, \tag{4.86}$$

where C_0 is a constant of integration. This Yukawa coupling takes a form different from the one we derived in Chap. 1. This is because we use here the deformation coordinate u which differs from the one used in Chap. 1. From a geometrical point of view, this u is more natural than ψ used in Chap. 1.

4.3.3 Instanton Expansion of A-Model Yukawa Couplings

In this subsection, we translate the B-model Yukawa couplings of $\widetilde{X}_{(\Delta^*,\Delta)}$ determined in the previous subsection into A-model Yukawa couplings of $\widetilde{X}_{(\Delta,\Delta^*)}$ under the hypothesis of mirror symmetry, and obtain the instanton expansion of these A-model Yukawa couplings. We discuss the example where $\widetilde{X}_{(\Delta,\Delta^*)}$ is a quintic hypersurface in CP^4. As the first step, we explicitly determine solutions of the differential equation:

$$\left(\left(\frac{d}{du}\right)^4 - 5e^u\prod_{j=1}^{4}\left(5\frac{d}{du}+j\right)\right)w(u) = 0, \tag{4.87}$$

which was used in the previous subsection. We use the Frobenius method, which is a standard way to solve linear differential equations with regular singular points. In this method, we start by assuming that the solution is written in the following form:

$$w(u;\varepsilon) = \sum_{n=0}^{\infty} a_n(\varepsilon)e^{(n+\varepsilon)u}. \tag{4.88}$$

By substituting (4.88) into (4.87), we obtain the following recursive formula for $a_n(\varepsilon)$:

$$\begin{aligned} a_n(\varepsilon) &= 5\frac{(5n-1+5\varepsilon)(5n-2+5\varepsilon)(5n-3+5\varepsilon)(5n-4+5\varepsilon)}{(n+\varepsilon)^4}a_{n-1}(\varepsilon) \\ &= \frac{\prod_{j=0}^{4}(5n-j+5\varepsilon)}{(n+\varepsilon)^5}a_{n-1}(\varepsilon) \qquad (n \geq 1). \end{aligned} \tag{4.89}$$

The solution of this recursive formula with the initial condition $a_0(\varepsilon) = 1$ is given by $a_n(\varepsilon) = \frac{\prod_{j=1}^{5n}(j+5\varepsilon)}{\prod_{j=1}^{n}(j+\varepsilon)^5}$. Hence we replace $w(u;\varepsilon)$ by

$$w(u;\varepsilon) = \sum_{n=0}^{\infty} \frac{\prod_{j=1}^{5n}(j+5\varepsilon)}{\prod_{j=1}^{n}(j+\varepsilon)^5} e^{(n+\varepsilon)u}, \tag{4.90}$$

and substitute it into (4.87) again. Then we obtain the following relation:

$$\left(\left(\frac{d}{du}\right)^4 - 5e^u \prod_{j=1}^{4}\left(5\frac{d}{du}+j\right)\right) w(u;\varepsilon) = \varepsilon^4 e^{\varepsilon u}. \tag{4.91}$$

By differentiating both sides of the above relation by ε and setting ε to 0, we can see that $w_j(u) := \frac{\partial^j w(u;\varepsilon)}{\partial \varepsilon^j}|_{\varepsilon=0}$ $(j = 0, 1, 2, 3)$ are solutions. The solution $w_j(u)$ has the following structure:

$$\begin{aligned}
w_j(u) &= \sum_{i=0}^{j} \binom{j}{i} u^{j-i} \widetilde{w}_j(u), \\
\widetilde{w}_j(u) &:= \sum_{n=0}^{\infty} \left(\frac{\partial^j}{\partial \varepsilon^j} \left(\frac{\prod_{j=1}^{5n}(j+5\varepsilon)}{\prod_{j=1}^{n}(j+\varepsilon)^5} \right)\Bigg|_{\varepsilon=0} \right) e^{nu}, \\
w_0(u) &= \widetilde{w}_0(u) = \sum_{n=0}^{\infty} \frac{(5n)!}{(n!)^5} e^{nu},
\end{aligned} \tag{4.92}$$

where $\widetilde{w}_j(u)$ is a power series in e^u. We can obtain a more explicit form of the expansion coefficients by executing differentiation by ε in the second line, but the result is quite complicated. In practice, we had better put the above formula into a computer algebra system to compute the explicit instanton expansion. Since $w_j(u)$ contains $u^j \cdot w_0(u)$ $(u = \log(e^u))$ as the top term in the expansion, it is called the j-th log solution.

According to the mirror symmetry hypothesis, the deformation parameter u of the complex structure of $\widetilde{X}_{(\Delta^*,\Delta)}$ is related to the deformation parameter t of the Kähler structure of $\widetilde{X}_{(\Delta,\Delta^*)}$ via the following mirror map, which is given by the first log solution divided by the power series solution:

$$\begin{aligned}
t = t(u) &= \frac{w_1(u)}{w_0(u)} = u + \frac{\widetilde{w}_1(u)}{\widetilde{w}_0(u)} \\
&= u + 770e^u + 717825e^{2u} + \frac{3225308000}{3}e^{3u} + \frac{3947314570625}{2}e^{4u} + \cdots \\
&=: u + \sum_{n=0}^{\infty} b_n e^{nu}.
\end{aligned} \tag{4.93}$$

The parameter t in the above equation is different from the one used in Chap. 1 up to multiplication of $2\pi i$, but we use the above definition for simplicity of computation. In order to obtain the instanton expansion of the A-model Yukawa coupling, we have to determine the expansion coefficients of the inverse function $u = u(t)$. For this purpose, we assume the following expansion form of the inverse function:

$$u = u(t) = t + \sum_{n=0}^{\infty} c_n e^{nt}. \tag{4.94}$$

We substitute (4.94) into (4.93) and compute the expansion form $t = t(u(t)) = t + \sum_{n=1}^{\infty} d_n e^{nt}$ by using the following relation:

$$e^{nu(t)} = e^{nt} \cdot \exp(n(\sum_{j=1}^{\infty} c_j e^{jt})) = e^{nt} \prod_{j=1}^{\infty} \exp(nc_j e^{jt}) = e^{nt} \prod_{j=1}^{\infty} \Big(\sum_{m_j=0}^{\infty} \frac{(nc_j e^{jt})^{m_j}}{m_j!} \Big). \tag{4.95}$$

Obviously the expansion coefficients d_n $(n \geq 1)$ should vanish. This condition leads us to recursive constraints on the unknown expansion coefficients c_n and enables us to determine them degree by degree. This operation is fundamentally simple, but in practice, we had better use computers because expansion coefficients are large numbers. Lower expansion coefficients of the inverse mirror map $u(t)$ are given as follows:

$$u(t) = t - 770e^t - 124925e^{2t} - \frac{305179250}{3} e^{3t} - \frac{198337448125}{2} e^{4t} - \cdots. \tag{4.96}$$

With these preparations, we translate Y_{uuu}, the B-model Yukawa coupling of $\widetilde{X}_{(\Delta^*,\Delta)}$ into $\langle \mathcal{O}_h(z_1)\mathcal{O}_h(z_2)\mathcal{O}_h(z_3)\rangle$, the A-model Yukawa coupling of $\widetilde{X}_{(\Delta,\Delta^*)}$. The rule of translation is given by

$$\begin{aligned}
\langle \mathcal{O}_h(z_1)\mathcal{O}_h(z_2)\mathcal{O}_h(z_3)\rangle &= \Big(\int_{\widetilde{X}_{(\Delta^*,\Delta)}} \Big(\frac{\Omega(u)}{w_0(u)}\Big) \wedge \partial_u^3 \Big(\frac{\Omega(u)}{w_0(u)}\Big)\Big) \Big(\frac{du}{dt}\Big)^3 \\
&= \frac{Y_{uuu}(u(t))}{(w_0(u(t)))^2} \Big(\frac{du(t)}{dt}\Big)^3 \\
&= \frac{C_0}{(w_0(u(t)))^2(1-5^5 e^{u(t)})} \cdot \Big(\frac{du(t)}{dt}\Big)^3.
\end{aligned} \tag{4.97}$$

The factor $\big(\frac{du(t)}{dt}\big)^3$ comes from the fact that the Yukawa coupling is a symmetric 3-form with respect to the deformation parameter. On the other hand, the period integral $\Pi_i(u)$'s are regarded as a kind of homogeneous coordinates of the moduli space, and we have to divide them by $w_0(u)$, which is also one of the homogeneous coordinates, to obtain the actual local coordinates of the moduli space. This is the reason why we divide $\Omega(u)$ by $w_0(u)$. We choose the power series solution $w_0(u)$

because the mirror map is given by $\frac{w_1(u)}{w_0(u)}$. At this stage, we have to determine the value of the constant of integration C_0. By expanding the last line of (4.97) in powers of e^t, we can observe that the top constant term equals C_0. But in the formula (3.81) of Chap. 3, we have derived that the top constant term of $\langle \mathcal{O}_h(z_1)\mathcal{O}_h(z_2)\mathcal{O}_h(z_3)\rangle$ is given by 5. Hence we have to set C_0 to 5. With these considerations, we finally obtain the instanton expansion of $\langle \mathcal{O}_h(z_1)\mathcal{O}_h(z_2)\mathcal{O}_h(z_3)\rangle$:

$$\begin{aligned}\langle \mathcal{O}_h(z_1)\mathcal{O}_h(z_2)\mathcal{O}_h(z_3)\rangle &= \frac{5}{(w_0(u(t)))^2(1-5^5 e^{u(t)})}\cdot\left(\frac{du(t)}{dt}\right)^3\\ &= 5 + 2875e^t + 4876875e^{2t} + 8564575000e^{3t} + 15517926796875e^{4t}\\ &+ 28663236110956000e^{5t} + 53621944306062201000e^{6t}\\ &+ 101216230345800061125625e^{7t} + \cdots . \end{aligned} \tag{4.98}$$

In practice, the computational process of expansion is very hard because of large expansion coefficients and we had better leave it to computers. As we have discussed in Chap. 3, the instanton expansion of $\langle \mathcal{O}_h(z_1)\mathcal{O}_h(z_2)\mathcal{O}_h(z_3)\rangle$ has the following structure:

$$\langle \mathcal{O}_h(z_1)\mathcal{O}_h(z_2)\mathcal{O}_h(z_3)\rangle = 5 + \sum_{d=1}^{\infty}\frac{d^3 n_d e^{dt}}{1-e^{dt}}, \tag{4.99}$$

where n_d is "the number of degree d rational curves" in a quintic hypersurface in CP^4. By comparing (4.98) with (4.99), we obtain

$$\begin{aligned} &n_1 = 2875,\ n_2 = 609250,\ n_3 = 317206375,\ n_4 = 242467530000,\\ &n_5 = 229305888887625,\ n_6 = 248249742118022000. \end{aligned} \tag{4.100}$$

In this way, we can observe that n_d turns out to be a positive integer. If we expand the r.h.s of (4.99) in powers of e^t, we can see that this result is quite non-trivial. From this phenomenon, we realize the striking impact of the mirror symmetry hypothesis.

References

1. V.V. Batyrev. *Dual polyhedra and mirror symmetry for Calabi–Yau hypersurfaces in toric varieties*. J. Algebraic Geom. 3 (1994), no. 3, 493–535.
2. D.A. Cox, S. Katz. *Mirror Symmetry and Algebraic Geometry*. Mathematical Surveys and Monographs, American Mathematical Society, 1999.
3. S. Hosono, A. Klemm, S. Theisen, S.-T. Yau. *Mirror symmetry, mirror map and applications to Calabi–Yau hypersurfaces*. Comm. Math. Phys. 167 (1995), no. 2, 301–350.

Chapter 5
Reconstruction of Mirror Symmetry Hypothesis from a Geometrical Point of View

Abstract In this chapter, we discuss degree d instanton correction of the A-model Yukawa coupling of a quintic hypersurface derived in the previous chapter, from the point of view of geometry of the moduli space of degree d holomorphic maps from CP^1 to CP^4. First, we naively compactify the moduli space into $CP^{5(d+1)-1}$ and compute the corresponding instanton correction. Next, we introduce a refined moduli space $\overline{Mp}_{0,2}(CP^4, d)$, which is the compactified moduli space of degree d polynomial maps with two marked points, and compute intersection numbers of the moduli space that are related to instanton corrections of the quintic hypersurface [8]. We show that generating functions of these intersection numbers reproduce the period integrals used in the B-model computation. Then we reconstruct the mirror formula of the A-model Yukawa coupling proposed in the previous chapter by using these generating functions. Lastly, we explain the geometrical meaning of the coordinate transformation induced by the mirror map $u = u(t)$. In some sense, discussions in this chapter correspond to mathematical justification of the mirror symmetry hypothesis in the case of a quintic hypersurface in CP^4, which was rigorously done in the celebrated works by Givental [1] and by Lian, Liu and Yau [2]. But we present here our new approach from the point of view of "naive compactification of moduli space".

5.1 Simple Compactification of Holomorphic Maps from CP^1 to CP^4

5.1.1 A-Model Correlation Functions as Intersection Numbers

Let M be a compact Kähler manifold. In Chap. 3, we introduced the degree d correlation function of the topological sigma model (A-model) (in this chapter, the term "topological sigma model" always means A-model),

M. Jinzenji, *Classical Mirror Symmetry*, SpringerBriefs in Mathematical Physics, https://doi.org/10.1007/978-981-13-0056-1_5

$$\langle \mathcal{O}_{W_1}(z_1)\mathcal{O}_{W_2}(z_2)\cdots\mathcal{O}_{W_n}(z_n)\rangle_d = \langle \prod_{I=1}^{n} \mathcal{O}_{W_I}(z_I)\rangle_d. \tag{5.1}$$

Let us remind the readers of its geometrical meaning. In Chap. 3, we discussed a subtle situation where non-trivial ψ-zero modes appear, but here, we assume that no ψ-zero mode appears. In (5.1), W_I is an element cohomology ring of M, $W_I \in H^{p_I,q_I}(M, \mathbf{C})$, and each W_I determines a codimension $p_I + q_I$ closed submanifold $PD(W_I)$ (up to homology equivalence). z_I's are distinct points on CP^1. Under this setting, the correlation function was characterized geometrically as follows:

- $\langle \prod_{I=1}^{n} \mathcal{O}_{W_I}(z_I)\rangle_d$ is the number of degree d holomorphic maps $\phi : CP^1 \to M$ that satisfy the condition:

$$\phi(z_I) \in PD(W_I), \ (I = 1, 2, \ldots, n). \tag{5.2}$$

We represent the degree d correlation function as an integral on the moduli space $\mathcal{M}_{CP^1}(M, d)$ of degree d holomorphic maps from CP^1 to M by using this characterization.

In this process, the notion of "compactification of the moduli space" becomes important. Representing the correlation function as an integral on the moduli space is based on intersection theory and the Poincaré dual, but these can be applied only to a compact, or closed space. In Chap. 3, we introduced the moduli space $\mathcal{M}_{CP^1}(CP^4, d)$, but as can be seen in the discussions in the chapter, it is an open space. Hence we need to compactify it by adding appropriate boundary components. Perhaps the most familiar example of compactification is compactification of the complex plane $\mathbf{C}$ into Riemann's sphere by adding the point of infinity $\{\infty\}$. To tell the truth, this method of compactifying a space is not unique, and in some cases, the correlation function takes different values for different compactifications. But here, we conventionally denote the compactified moduli space by $\overline{\mathcal{M}}_{CP^1}(M, d)$.

In this section, we use "simple compactification", which uses the complex projective space $CP^{5(d+1)-1}$ as $\overline{\mathcal{M}}_{CP^1}(CP^4, d)$. This idea was already mentioned in Chap. 3. Let us remind the readers of the formula (3.63) presented in Sect. 3.2.4:

$$CP^{5(d+1)-1} - \mathcal{M}_{CP^1}(CP^4, d) = \coprod_{f=1}^{d} \big(CP^f \times \mathcal{M}_{CP^1}(CP^4, d-f)\big). \tag{5.3}$$

As we have already discussed, $\dim_{\mathbf{C}}(CP^{5(d+1)-1})$ equals $\dim_{\mathbf{C}}(\mathcal{M}_{CP^1}(CP^4, d)) = 5d+4$. On the other hand, $\dim_{\mathbf{C}}(CP^f \times \mathcal{M}_{CP^1}(CP^4, d-f))$ is given by $f + 5(d-f) + 4 = 5d + 4 - 4f$ and less than $5d + 4$. Hence the compact space $CP^{5(d+1)-1}$ is obtained from $\mathcal{M}_{CP^1}(CP^4, d)$ by adding boundary components of positive codimension.

Now, we return to representing the correlation function as an integral on the moduli space. In order to rewrite the condition (5.2) in terms of closed differential forms on the moduli space, we define the evaluation map $ev_I : \overline{\mathcal{M}}_{CP^1}(M, d) \to M$ that evaluates the value of the map $\phi \in \overline{\mathcal{M}}_{CP^1}(M, d)$ at $z_I \in CP^1$:

$$ev_I(\phi) = \phi(z_I) \in M. \tag{5.4}$$

Then the pull-back of $W_I \in H^{p_I, q_I}(M, \mathbf{C})$ by ev_I, $ev_I^*(W_I)$, becomes a closed (p_I, q_I) differential form on $\overline{\mathcal{M}}_{CP^1}(M, d)$, i.e., an element of $H^{p_I, q_I}(\overline{\mathcal{M}}_{CP^1}(M, d), \mathbf{C})$. Since $\overline{\mathcal{M}}_{CP^1}(M, d)$ is a compact space, it determines $PD(ev_I^*(W_I))$, which is a codimension $p_I + q_I$ closed subspace in $\overline{\mathcal{M}}_{CP^1}(M, d)$. Taking geometrical meaning of pull-back by the evaluation map into account, we obtain the following relation:

$$PD(ev_I^*(W_I)) = \{\phi \in \overline{\mathcal{M}}_{CP^1}(M, d) \mid \phi(z_I) \in PD(W_I)\}. \tag{5.5}$$

Then let us assume that the topological selection rule introduced in Chap. 3,

$$\sum_{I=1}^{n} p_I = \sum_{I=1}^{n} q_I = \dim_{\mathbf{C}}(\overline{\mathcal{M}}_{CP^1}(M, d)), \tag{5.6}$$

holds. By (5.5), the set of holomorphic maps that satisfy (5.2) is represented as

$$\bigcap_{i=1}^{n} PD(ev_I^*(W_I)), \tag{5.7}$$

and its codimension in $\overline{\mathcal{M}}_{CP^1}(M, d)$ equals $\sum_{I=1}^{n}(p_I + q_I) = 2\dim_{\mathbf{C}}(\overline{\mathcal{M}}_{CP^1}(M, d))$. Hence its real dimension is given by $2\dim_{\mathbf{C}}(\overline{\mathcal{M}}_{CP^1}(M, d)) - 2\dim_{\mathbf{C}}(\overline{\mathcal{M}}_{CP^1}(M, d)) = 0$ and the set turns out to be a discrete set of points. Hence we obtain the following relation:

$$\begin{aligned}\langle\prod_{I=1}^{n} \mathcal{O}_{W_I}(z_I)\rangle_d &= (\text{number of points in } \bigcap_{i=1}^{n} PD(ev_I^*(W_I))) \\ &=: \left(\bigcap_{i=1}^{n} PD(ev_I^*(W_I))\right)^{\sharp}.\end{aligned} \tag{5.8}$$

In this way, the correlation function is rewritten into an intersection number of $\overline{\mathcal{M}}_{CP^1}(M, d)$. According to the intersection theory on a compact space, intersection numbers of closed subspaces represented as Poincaré duals of closed differential forms are translated into an integral of the wedge product of corresponding closed differential forms. Hence (5.8) is further rewritten as follows:

$$\langle\prod_{I=1}^{n} \mathcal{O}_{W_I}(z_I)\rangle_d = \left(\bigcap_{i=1}^{n} PD(ev_I^*(W_I))\right)^{\sharp} = \int_{\overline{\mathcal{M}}_{CP^1}(M, d)} \bigwedge_{i=1}^{n} ev_I^*(W_I). \tag{5.9}$$

This is the formula we are aiming at.

5.1.2 Evaluation of Yukawa Coupling of Quintic Hypersurface in CP^4 by Using Simple Compactification of the Moduli Space

In this subsection, we compute the degree d correlation function $\langle \mathcal{O}_h(z_1)\mathcal{O}_h(z_1)\mathcal{O}_h(z_3)\rangle_d$ of a quintic hypersurface in CP^4, which corresponds to degree d instanton correction of the Yukawa coupling, by using the simple compactification $\overline{\mathcal{M}}_{CP^1}(CP^4, d) = CP^{5(d+1)-1}$ introduced in the previous subsection. From now on, we denote quintic hypersurface in CP^4 by M_4^5. As was discussed in Chap. 3, we have the subtlety of ψ-zero modes, which arises when the degree d holomorphic map ϕ is given by composition of a map $m_f : CP^1 \to CP^1$ of degree f $(f > 1,\ f|d)$ and a map $\tilde{\phi} : CP^1 \to M_4^5$ of degree $\frac{d}{f}$. But in this subsection, we proceed by conveniently regarding the degree d correlation function $\langle \mathcal{O}_h(z_1)\mathcal{O}_h(z_1)\mathcal{O}_h(z_3)\rangle_d$ as the number of holomorphic maps $\phi : CP^1 \to CP^4$ that satisfy the following conditions:

(i) $\phi(z_i) \in PD(h)$ $(i = 1, 2, 3)$,
(ii) $\phi(CP^1) \subset M_4^5$.

Let us consider the condition (i) first. The cohomology ring of $CP^{5(d+1)-1}$ is generated by the hyperplane class $\tilde{h} \in H^{1,1}(CP^{5(d+1)-1})$. $\tilde{h}$ satisfies the relation $(\tilde{h})^{5(d+1)} = 0$ and the normalization condition:

$$\int_{CP^{5(d+1)-1}} (\tilde{h})^{5(d+1)-1} = 1. \tag{5.10}$$

On the other hand, a point in $CP^{5(d+1)-1}$ is obtained by representing a rational map ϕ from CP^1 to CP^4 in the form

$$\phi(s,t) = (\sum_{j=0}^{d} a_j^1 s^j t^{d-j} : \sum_{j=0}^{d} a_j^2 s^j t^{d-j} : \cdots : \sum_{j=0}^{d} a_j^5 s^j t^{d-j}), \tag{5.11}$$

and by regarding its coefficients a_j^i as homogeneous coordinates:

$$(a_0^1 : a_1^1 : a_2^1 : \cdots : a_{d-1}^5 : a_d^5) \in CP^{(5(d+1)-1}. \tag{5.12}$$

Let $PD(h)$ be a hyperplane $\{\ (X_1 : \cdots : X_5) \in CP^4 \mid X_5 = 0\ \}$ for example. By setting $z_i = (s_i : t_i)$, the condition $\phi(z_i) \in PD(h)$ leads us to the following equality:

$$\sum_{j=0}^{d} a_j^5 (s_i)^j (t_i)^{d-j} = 0. \tag{5.13}$$

Since s_i and t_i are constants, the subset that satisfies the above condition is nothing but a hyperplane in $CP^{5(d+1)-1}$. Therefore, the condition $\phi(z_i) \in PD(h)$ is translated into the insertion of $\tilde{h}$. Therefore, the condition (i) is realized by inserting $\tilde{h}^3 \in H^*(CP^{5(d+1)-1}, \mathbf{C})$ into the integration of differential forms on $CP^{5(d+1)-1}$ by using the discussions in the previous subsection. As for the condition (ii), we use the discussions in Chap. 3. We take the defining equation of M_4^5 in the form,

$$(X_1)^5 + (X_2)^5 + (X_3)^5 + (X_4)^5 + (X_5)^5 = 0, \tag{5.14}$$

for simplicity. Then the condition $\phi(CP^1) \in M_4^5$ is translated into the condition that the following equation:

$$(\sum_{j=0}^{d} a_j^1 s^j t^{d-j})^5 + (\sum_{j=0}^{d} a_j^2 s^j t^{d-j})^5 + \cdots + (\sum_{j=0}^{d} a_j^5 s^j t^{d-j})^5 = 0 \Longleftrightarrow \sum_{m=0}^{5d} f_m^5(a_j^i) s^m t^{5d-m} = 0, \tag{5.15}$$

holds for any $(s:t) \in CP^1$. Here, $f_m^5(a_j^i)$ is defined as the coefficient of $s^m t^{5d-m}$ of the polynomial obtained from expanding the l.h.s. of the first line, and it is a degree 5 homogeneous polynomial in a_j^i. Since $(s:t)$ is arbitrary, the above condition is equivalent to

$$f_m(a_j^i) = 0, \quad (m = 0, 1, 2, \ldots, 5d). \tag{5.16}$$

Let us explain how the condition (5.16) is translated into the language of the cohomology ring of $CP^{5(d+1)-1}$. First, we note that $f_m^5(a_j^i)$ is a degree 5 homogeneous polynomial. As was discussed in Chap. 2, this means that it is a global section of $\tilde{H}^{\otimes 5} = \mathcal{O}_{CP^{5(d+1)-1}}(5)$, which is a line bundle obtained as the 5-times tensor product of the hyperplane bundle $\tilde{H}$ on $CP^{5(d+1)-1}$. At this stage, we arrange $f_m^5(a_j^i)$'s into the following form:

$$F_d^5 := (f_0^5(a_j^i), f_2^5(a_j^i), \cdots, f_{5d}^5(a_j^i)). \tag{5.17}$$

Then F_d^5 can be regarded as a global section of a rank $5d+1$ holomorphic vector bundle on $CP^{5(d+1)-1}$,

$$\begin{aligned} E_d^5 &= \overbrace{\mathcal{O}_{CP^{5(d+1)-1}}(5) \oplus \mathcal{O}_{CP^{5(d+1)-1}}(5) \oplus \cdots \oplus \mathcal{O}_{CP^{5(d+1)-1}}(5)}^{5d+1} \\ &= \mathcal{O}_{CP^{5(d+1)-1}}(5)^{\oplus 5d+1}. \end{aligned} \tag{5.18}$$

With this set-up, the condition (5.16) is translated into the condition that ϕ belongs to the zero locus of the global section F_d^5 of the vector bundle E_d^5. Then we introduce a theorem in mathematics called the "generalized Gauss–Bonnet theorem".

Theorem 5.1.1 *Let M be a compact complex manifold and E be a rank r holomorphic vector bundle on M. Then Poincaré dual of zero locus of a holomorphic global section of E is given by the r-th Chern class of E.*

This translates the condition (5.16) into the insertion of $c_{5d+1}(E_d^5) = c_{5d+1}(\mathcal{O}_{CP^{5(d+1)-1}}(5)^{\oplus 5d+1})$ in the integration on $CP^{5(d+1)-1}$. By using the results in

Chap. 2, we can compute the total Chern class $c(E_d^5)$ of E_d^5 as follows:

$$\begin{aligned} c(E_d^5) &= c(\mathcal{O}_{CP^{5(d+1)-1}}(5)^{\oplus 5d+1}) \\ &= c(\mathcal{O}_{CP^{5(d+1)-1}}(5))^{5d+1} = (1+t(5\tilde{h}))^{5d+1}. \end{aligned} \tag{5.19}$$

Hence we obtain $c_{5d+1}(E_d^5) = (5\tilde{h})^{5d+1}$. Combining the results obtained so far, the degree d instanton correction of M_4^5 under the compactification $\overline{\mathcal{M}}_{CP^1}(CP^4, d) = CP^{5(d+1)-1}$ is computed as follows:

$$\begin{aligned} \langle \mathcal{O}_h(z_1)\mathcal{O}_h(z_1)\mathcal{O}_h(z_3)\rangle_d &= \int_{CP^{5(d+1)-1}} \tilde{h}^3 \wedge c_{5d+1}(E_d^5) \\ &= \int_{CP^{5(d+1)-1}} \tilde{h}^3 \wedge (5\tilde{h})^{5d+1} \\ &= 5^{5d+1}\int_{CP^{5(d+1)-1}} \tilde{h}^{5d+4} = 5^{5d+1}. \end{aligned} \tag{5.20}$$

Unfortunately, this result does not coincide with the prediction of the mirror symmetry hypothesis except for the $d = 0$ case. For example, in the $d = 1$ case, the mirror prediction (4.98) tells us

$$\langle \mathcal{O}_h(z_1)\mathcal{O}_h(z_1)\mathcal{O}_h(z_3)\rangle_1 = 2875, \tag{5.21}$$

but (5.20) gives us $5^6 = 15625$, which is greater than 2875. Also in $d > 1$ cases, 5^{5d+1} is greater than the mirror prediction given by (4.98). This is because contributions from boundary components added in compactifying $\mathcal{M}_{CP^1}(CP^4, d)$ into $\overline{\mathcal{M}}_{CP^1}(CP^4, d)$ cannot be neglected in integration. Let us explain the situation briefly.

As was discussed in Sect. 3.2.4, a point ϕ in the boundary component $CP^f \times \mathcal{M}_{CP^1}(CP^4, d-f)$ is described by a polynomial map each of whose degree d homogeneous polynomial $\sum_{j=0}^{d} a_j^i s^j t^{d-j}$ is factorized into the following form:

$$\sum_{j=0}^{d} a_j^i s^j t^{d-j} = \left(\sum_{j=0}^{f} b_j s^j t^{f-j}\right)\cdot\left(\sum_{j=0}^{d-f} c_j^i s^j t^{d-f-j}\right), \quad (i = 1, \ldots, 5), \tag{5.22}$$

where $\sum_{j=0}^{f} b_j s^j t^{f-j}$ is the common divisor of the highest degree (hence $\sum_{j=0}^{d-f} c_j^i s^j t^{d-f-j}$ $(i = 1, \ldots, 5)$ do not have a common divisor of positive degree). Then the condition $\phi(CP^1) \subset M_4^5$ given by (5.15) becomes equivalent to the condition that the equation:

$$
\begin{aligned}
&(\sum_{j=0}^{f} b_j s^j t^{f-j})^5 \cdot (\sum_{j=0}^{d-f} c_j^1 s^j t^{d-f-j})^5 + \cdots + (\sum_{j=0}^{f} b_j s^j t^{f-j})^5 \cdot (\sum_{j=0}^{d-f} c_j^5 s^j t^{d-f-j})^5 = 0 \\
&\Longleftrightarrow (\sum_{j=0}^{f} b_j s^j t^{f-j})^5 \cdot (\sum_{m=0}^{5(d-f)} f_m^5(c_j^i) s^m t^{5(d-f)-m}) = 0,
\end{aligned} \tag{5.23}
$$

holds for any $(s:t) \in CP^1$. If $(\sum_{j=0}^{f} b_j s^j t^{f-j})^5 = 0$ holds for any $(s:t)$, it follows that $b_0 = b_1 = \cdots = b_f = 0$. Hence it is impossible. Therefore, the above condition reduces to the condition that

$$
\sum_{m=0}^{5(d-f)} f_m^5(c_j^i) s^m t^{5(d-f)-m} = 0, \tag{5.24}
$$

holds for any $(s:t)$, i.e.,

$$
f_m^5(c_j^i) = 0, \quad (m = 0, 1, \ldots, 5(d-f)). \tag{5.25}
$$

Even if these equations are independent, they reduce the dimension of the moduli space at most by $5(d-f)+1$. Moreover, if the condition (5.25) holds, (5.23) is satisfied automatically and the parameters $(b_0 : b_1 : \cdots : b_f) \in CP^f$ can vary freely. Hence the complex dimension of the set of points in the boundary component $CP^f \times \mathcal{M}_{CP^1}(CP^4, d-f)$ that satisfy the condition (5.23) is no less than $f + 5((d-f)+1) - 1 - (5(d-f)+1) = f + 3 (>3)$. This dimension is greater than 3, which is the complex dimension of the set of points in $\mathcal{M}_{CP^1}(CP^4, d)$ satisfying the condition (5.23). In this situation, we are forced to admit that the result 5^{5d+1} obtained in (5.20) contains unwanted contributions from boundary components. This is the reason why it is greater than the degree d instanton correction obtained from the mirror symmetry hypothesis. Therefore, in order to obtain the correct degree d instanton correction, we have to "subtract" unwanted contributions from the boundary components.

On the other hand, the result given in (5.20) has a surprising connection with the process of B-model computation. Let us assume that the result (5.20) is the correct instanton correction $\langle \mathcal{O}_h(z_1)\mathcal{O}_h(z_1)\mathcal{O}_h(z_3)\rangle_d$ and compute the generating function of $\langle \mathcal{O}_h(z_1)\mathcal{O}_h(z_1)\mathcal{O}_h(z_3)\rangle_d$ by using e^u as a formal variable.

$$
\sum_{d=0}^{\infty} \langle \mathcal{O}_h(z_1)\mathcal{O}_h(z_1)\mathcal{O}_h(z_3)\rangle_d e^{du} = \sum_{d=0}^{\infty} 5^{5d+1} e^{du} = \frac{5}{1-5^5 e^u}. \tag{5.26}
$$

This result reproduces the B-model Yukawa coupling Y_{uuu} in (4.86) with $C_0 = 5$! In the final stage of B-model computation, C_0 was adjusted to 5. Hence it gives the correct B-model Yukawa coupling. This phenomenon seems to imply a deep connection between B-model computation and the simple compactification of the moduli space.

5.2 Toric Compactification of the Moduli Space of Degree d Quasi Maps with Two Marked Points

In the previous section, we used the following representation of the holomorophic map from CP^1 to CP^4 of degree d:

$$\phi(s,t) = (\sum_{j=0}^{d} a_j^1 s^j t^{d-j} : \sum_{j=0}^{d} a_j^2 s^j t^{d-j} : \cdots : \sum_{j=0}^{d} a_j^5 s^j t^{d-j}). \tag{5.27}$$

In this section, we rewrite the above expression by introducing a polynomial map $p : \mathbf{C}^2 \to \mathbf{C}^5$:

$$p(s,t) = \mathbf{a}_0 s^d + \mathbf{a}_1 s^{d-1} t + \mathbf{a}_2 s^{d-2} t^2 + \cdots + \mathbf{a}_d t^d, \tag{5.28}$$

where $\mathbf{a}_j = (a_j^1, a_j^2, \ldots, a_j^5) \in \mathbf{C}^5$. The holomorphic map ϕ from CP^1 to CP^4 is recovered as $\phi(s:t) = \pi_5(p(s,t))$ where $\pi_5 : \mathbf{C}^5 \setminus \{\mathbf{0}\} \to CP^4$ is the natural projection. The parameter space of the above polynomial maps is given by $\mathbf{C}^{5(d+1)} = \{(\mathbf{a}_0, \mathbf{a}_1, \ldots, \mathbf{a}_d)\}$. We denote by $Mp_{0,2}(CP^4, d)$ the space obtained from dividing $\{(\mathbf{a}_0, \ldots, \mathbf{a}_d) \in \mathbf{C}^{5(d+1)} |\ \mathbf{a}_0 \neq \mathbf{0},\ \mathbf{a}_d \neq \mathbf{0}\}$ by two $\mathbf{C}^{\times}$ actions induced from the following two $\mathbf{C}^{\times}$ actions on $\mathbf{C}^2$ via the map p in (5.28):

$$(s,t) \to (\mu s, \mu t), \quad (s,t) \to (s, \nu t). \tag{5.29}$$

The second $\mathbf{C}^{\times}$ action corresponds to the automorphism group of CP^1 that keeps $0(= (1:0))$ and $\infty(= (0:1))$ fixed. With the above two torus actions, $Mp_{0,2}(CP^4, d)$ can be regarded as the parameter space of degree d polynomial maps from CP^1 to CP^4 with two marked points in CP^1: $0(= (1:0))$ and $\infty(= (0:1))$. Set theoretically, it is given as follows:

$$Mp_{0,2}(CP^4, d) = \{(\mathbf{a}_0, \mathbf{a}_1, \ldots, \mathbf{a}_d) \in \mathbf{C}^{5(d+1)} \mid \mathbf{a}_0, \mathbf{a}_d \neq \mathbf{0}\}/(\mathbf{C}^{\times})^2, \tag{5.30}$$

where the two $\mathbf{C}^{\times}$ actions are given by

$$\begin{aligned}(\mathbf{a}_0, \mathbf{a}_1, \ldots, \mathbf{a}_d) &\to (\mu\mathbf{a}_0, \mu\mathbf{a}_1, \ldots, \mu\mathbf{a}_{d-1}, \mu\mathbf{a}_d)\\ (\mathbf{a}_0, \mathbf{a}_1, \ldots, \mathbf{a}_d) &\to (\mathbf{a}_0, \nu\mathbf{a}_1, \ldots, \nu^{d-1}\mathbf{a}_{d-1}, \nu^d\mathbf{a}_d)\end{aligned} \tag{5.31}$$

The condition $\mathbf{a}_0, \mathbf{a}_d \neq \mathbf{0}$ assures that the images of 0 and ∞ are well-defined in CP^4.

At this stage, we have to note the difference between the moduli space of holomorphic maps from CP^1 to CP^4 and the moduli space of polynomial maps from CP^1 to CP^4. In short, the latter includes the points that are not actual maps from

CP^1 to CP^4 but rational maps from CP^1 to CP^4. Let us consider a polynomial map $\sum_{j=0}^{d} \mathbf{a}_j s^j t^{d-j}$, which can be factorized as

$$\sum_{j=0}^{d} \mathbf{a}_j s^j t^{d-j} = p_{d-d_1}(s,t) \cdot (\sum_{j=0}^{d_1} \mathbf{b}_j s^j t^{d_1-j}), \tag{5.32}$$

where $p_{d-d_1}(s,t)$ is a homogeneous polynomial of degree $d - d_1(>0)$. If we consider $\sum_{j=0}^{d} \mathbf{a}_j s^j t^{d-j}$ as a map from CP^1 to CP^4, it should be regarded as a rational map whose image of the zero points of p_{d-d_1} is undefined. Moreover, the closure of the image of this map is a rational curve of degree $d_1(<d)$ in CP^4. The reason why we include these rational maps is that we can obtain simpler compactification of the moduli space, as was mentioned in the previous section.

Now, let us turn to the problem of compactification of $Mp_{0,2}(CP^4, d)$. If $d = 1$, $Mp_{0,2}(CP^4, 1)$ is given by

$$Mp_{0,2}(CP^4, 1) = \{(\mathbf{a}_0, \mathbf{a}_1) \in \mathbf{C}^{10} \mid \mathbf{a}_0, \mathbf{a}_1 \neq \mathbf{0}\}/(\mathbf{C}^\times)^2, \tag{5.33}$$

where $(\mathbf{C}^\times)^2$ action is given as follows:

$$\begin{aligned} (\mathbf{a}_0, \mathbf{a}_1) &\to (\mu\mathbf{a}_0, \mu\mathbf{a}_1) \\ (\mathbf{a}_0, \mathbf{a}_1) &\to (\mathbf{a}_0, \nu\mathbf{a}_1). \end{aligned} \tag{5.34}$$

Therefore, $Mp_{0,2}(CP^4, 1)$ is nothing but $CP^4 \times CP^4$ and is already compact. If $d \geq 2$, we have to use the two $\mathbf{C}^\times$ actions in (5.31) to turn $\mathbf{a}_0$ and $\mathbf{a}_d$ into the points in CP^4, $[\mathbf{a}_0]$ and $[\mathbf{a}_d]$, where $[\mathbf{a}] = \pi_5(\mathbf{a})$. Therefore, we can easily see that

$$Mp_{0,2}(CP^4, d) = \{([\mathbf{a}_0], \ldots, \mathbf{a}_{d-1}, [\mathbf{a}_d]) \in CP^4 \times \mathbf{C}^{5(d-1)} \times CP^4 \mid\}/\mathbf{Z}_d. \tag{5.35}$$

In (5.35), the $\mathbf{Z}_d$ acts on $\mathbf{C}^{5(d-1)}$ as follows:

$$(\mathbf{a}_1, \mathbf{a}_2 \cdots, \mathbf{a}_{d-1}) \to ((\zeta_d)^j \mathbf{a}_1, (\zeta_d)^{2j} \mathbf{a}_2 \ldots, (\zeta_d)^{(d-1)j} \mathbf{a}_{d-1}), \tag{5.36}$$

where ζ_d is the d-th primitive root of unity. In this way, we can see that $Mp_{0,2}(CP^4, d)$ is not compact if $d \geq 2$. In order to compactify $Mp_{0,2}(CP^4, d)$, we introduce the following chains of polynomial maps:

$$\cup_{j=1}^{l(\sigma_d)} \Big(\sum_{m_j=0}^{d_j-d_{j-1}} \mathbf{a}_{d_{j-1}+m_j} (s_j)^{m_j} (t_j)^{d_j-d_{j-1}-m_j}\Big), \quad (\mathbf{a}_{d_j} \neq \mathbf{0}, \ \ j = 0, 1, \ldots, l(\sigma_d)), \tag{5.37}$$

at the infinity locus of $Mp_{0,2}(CP^4, d)$. In (5.37), d_j's are integers that satisfy

$$1 \leq d_1 < d_2 < \cdots < d_{l(\sigma_d)} \leq d - 1. \tag{5.38}$$

We denote by $\overline{Mp}_{0,2}(CP^4, d)$ the space obtained after this compactification. This $\overline{Mp}_{0,2}(CP^4, d)$ is the moduli space we use in the remaining part of this book.[1] It is explicitly constructed as a toric orbifold by introducing boundary divisor coordinates $u_1, u_2, \ldots u_{d-1}$ as follows:

$$\overline{Mp}_{0,2}(CP^4, d) = \{(\mathbf{a}_0, \ldots, \mathbf{a}_d, u_1, \ldots, u_{d-1}) \in \mathbf{C}^{5(d+1)+d-1} \mid \mathbf{a}_0, (\mathbf{a}_1, u_1), \ldots, (\mathbf{a}_{d-1}, u_{d-1}), \mathbf{a}_d \neq \mathbf{0}\}/(\mathbf{C}^\times)^{d+1}, \quad (5.39)$$

where the $(d+1)$ $\mathbf{C}^\times$ actions are given by

$$\begin{aligned}
&(\mathbf{a}_0, \mathbf{a}_1, \ldots, \mathbf{a}_d, u_1, \ldots, u_{d-1}) \to (\mu_0\mathbf{a}_0, \ldots, \mu_0^{-1}u_1, \ldots),\\
&(\mathbf{a}_0, \mathbf{a}_1, \ldots, \mathbf{a}_d, u_1, \ldots, u_{d-1}) \to (\cdots, \mu_1\mathbf{a}_1, \ldots, \mu_1^2u_1, \mu_1^{-1}u_2, \ldots),\\
&(\mathbf{a}_0, \mathbf{a}_1, \ldots, \mathbf{a}_d, u_1, \ldots, u_{d-1}) \to (\cdots, \mu_i\mathbf{a}_i, \ldots, \mu_i^{-1}u_{i-1}, \mu_i^2u_i, \mu_i^{-1}u_{i+1}, \ldots)\\
&(i = 2, \ldots, d-1),\\
&(\mathbf{a}_0, \mathbf{a}_1, \ldots, \mathbf{a}_d, u_1, \ldots, u_{d-1}) \to (\cdots, \mu_{d-1}\mathbf{a}_{d-1}, \ldots, \mu_{d-1}^{-1}u_{d-2}, \mu_{d-1}^2u_{d-1}),\\
&(\mathbf{a}_0, \mathbf{a}_1, \ldots, \mathbf{a}_d, u_1, \ldots, u_{d-1}) \to (\cdots, \mu_d\mathbf{a}_d, \ldots, \mu_d^{-1}u_{d-1}). \qquad (5.40)
\end{aligned}$$

In (5.40), "$\cdots$" in the r.h.s indicates that the $\mathbf{C}^\times$ actions are trivial. These torus actions are represented by a $(d+1) \times 2d$ weight matrix W_d:

$$W_d := \begin{array}{c} \\ h_0 \\ h_1 \\ h_2 \\ \vdots \\ \vdots \\ \vdots \\ h_d \end{array} \begin{array}{c} \begin{array}{cccccccccccc} \mathbf{a}_0 & \mathbf{a}_1 & \mathbf{a}_2 & \cdots & \cdots & \mathbf{a}_{d-1} & \mathbf{a}_d & u_1 & u_2 & u_3 & \cdots & u_{d-1} \end{array} \\ \begin{pmatrix} 1 & 0 & 0 & \cdots & \cdots & 0 & 0 & -1 & 0 & 0 & \cdots & 0 \\ 0 & 1 & 0 & \cdots & \cdots & 0 & 0 & 2 & -1 & 0 & \cdots & 0 \\ 0 & 0 & 1 & \ddots & \cdots & \vdots & 0 & -1 & 2 & -1 & \ddots & 0 \\ \vdots & \vdots & \ddots & \ddots & \ddots & \vdots & \vdots & \vdots & \ddots & \ddots & \ddots & \vdots \\ \vdots & \vdots & \ddots & \ddots & \ddots & 0 & 0 & 0 & 0 & \ddots & \ddots & -1 \\ \vdots & \vdots & \vdots & \ddots & \ddots & 1 & 0 & 0 & 0 & \ddots & -1 & 2 \\ 0 & 0 & 0 & \cdots & \cdots & 0 & 1 & 0 & 0 & \cdots & 0 & -1 \end{pmatrix} \end{array}$$

Notice that the A_{d-1} Cartan matrix appears in W_d. If $u_1, u_2, \ldots, u_{d-1} \neq 0$, we can set all the u_i's to 1 by using the $(d+1)$ torus actions. The remaining two torus actions that leave them invariant are nothing but the ones given in (5.31). Therefore, the subspace given by the condition $u_1, u_2, \ldots, u_{d-1} \neq 0$ corresponds to $Mp_{0,2}(CP^4, d)$. If $u_{d_1} = 0,\ u_j \neq 0\ \ (j \neq d_1)$, we have to delete the u_{d_1} column of matrix W_d. This operation turns the A_{d-1} Cartan matrix into the $A_{d_1-1} \times A_{d-d_1-1}$ Cartan matrix and results in chains of two quasi maps:

[1]The moduli space $\overline{Mp}_{0,2}(CP^4, d)$ was also independently constructed by Ciocan-Fontanine and Kim [3] from the point of view of the stability of quasimaps.

$$(\sum_{j=0}^{d_1} \mathbf{a}_j s_1^j t_1^{d_1-j}) \cup (\sum_{j=0}^{d-d_1} \mathbf{a}_{j+d_1} s_2^j t_2^{d-d_1-j}), \quad (\mathbf{a}_0, \mathbf{a}_{d_1}, \mathbf{a}_d \neq \mathbf{0}). \tag{5.41}$$

Therefore, the corresponding boundary locus is given by $Mp_{0,2}(CP^4, d_1) \underset{CP^4}{\times} Mp_{0,2}(N, d-d_1)$, where $\underset{CP^{N-1}}{\times}$ is the fiber product with respect to the following projection maps:

$$\begin{aligned} &\pi_1 : Mp_{0,2}(CP^4, d_1) \to CP^4, \quad \pi_1(\mathbf{a}_0, \ldots, \mathbf{a}_{d_1}) = [\mathbf{a}_{d_1}], \\ &\pi_2 : Mp_{0,2}(CP^4, d-d_1) \to CP^4, \quad \pi_2(\mathbf{a}_{d_1}, \ldots, \mathbf{a}_d) = [\mathbf{a}_{d_1}]. \end{aligned} \tag{5.42}$$

In general, the subspace given by the condition

$$u_{d_i} = 0 \ (1 \leq d_1 < d_2 < \cdots < d_{l(\sigma_d)-1} \leq d-1), \, u_j \neq 0 \ (j \notin \{d_1, d_2, \ldots, d_{l(\sigma_d)-1}\}), \tag{5.43}$$

corresponds to chains of quasi maps labeled by the ordered partition $\sigma_d = (d_1 - d_0, d_2 - d_1, d_3 - d_2, \ldots, d_{l(\sigma_d)} - d_{l(\sigma_d)-1})$:

$$\cup_{j=1}^{l(\sigma_d)} \Big(\sum_{m_j=0}^{d_j-d_{j-1}} \mathbf{a}_{d_{j-1}+m_j} (s_j)^{m_j} (t_j)^{d_j-d_{j-1}-m_j}\Big) \ \big(\mathbf{a}_{d_j} \neq \mathbf{0}, \ j = 0, 1, \ldots, l(\sigma_d)\big), \tag{5.44}$$

where we set $d_0 = 0, d_{l(\sigma_d)} = d$. In this case, the corresponding boundary locus is

$$Mp_{0,2}(CP^4, d_1 - d_0) \underset{CP^4}{\times} Mp_{0,2}(CP^4, d_2 - d_1) \underset{CP^4}{\times} \cdots \underset{CP^4}{\times} Mp_{0,2}(CP^4, d_{l(\sigma_d)} - d_{l(\sigma_d)-1}). \tag{5.45}$$

Since the lowest dimensional boundary:

$$Mp_{0,2}(CP^4, 1) \underset{CP^4}{\times} Mp_{0,2}(CP^4, 1) \underset{CP^4}{\times} \cdots \underset{CP^4}{\times} Mp_{0,2}(CP^4, 1), \tag{5.46}$$

is identified with the compact space $(CP^4)^{d+1}$, we can conclude that $\overline{Mp}_{0,2}(CP^4, d)$ is compact.

Next, we discuss the structure of the cohomology ring (to be precise, Chow ring) $H^*(\overline{Mp}_{0,2}(CP^4, d))$. In (5.41), we labeled row vectors of W_d by h_i $(i = 0, 1, \ldots, d)$, which represents Kähler forms of $\overline{Mp}_{0,2}(CP^4, d)$ associated with the torus action of μ_i in (5.40). By using standard results on toric varieties, we can see that these h_i's are generators of $H^*(\overline{Mp}_{0,2}(CP^4, d))$ and that relations between the generators are given by the data of elements of W_d as follows:

$$\begin{aligned} &(h_0)^5 = 0, \ (h_d)^5 = 0, \\ &(h_i)^5(2h_i - h_{i-1} - h_{i+1}) = 0 \ \ (i = 1, 2, \ldots, d-1). \end{aligned} \tag{5.47}$$

Explicit proof of the above relations is given in [4].

5.3 Construction of Two Point Intersection Numbers on $\overline{Mp_{0,2}}(CP^4, d)$

In this section, we define the following intersection number on $\overline{Mp_{0,2}}(N, d)$, which is related to the correlation function of the A-model on a quintic hypersurface in CP^4:

$$w(\mathcal{O}_{h^a}\mathcal{O}_{h^b})_{0,d} := \int_{\overline{Mp_{0,2}}(CP^4,d)} ev_1^*(h^a) \wedge ev_2^*(h^b) \wedge c_{top}(\mathcal{E}_d). \tag{5.48}$$

In (5.48), h is the hyperplane class of CP^4, and $ev_1 : \overline{Mp_{0,2}}(CP^4, d) \to CP^4$ (resp. $ev_2 : \overline{Mp_{0,2}}(CP^4, d) \to CP^4$) is the evaluation map at the first (resp. second) marked point. These maps are easily constructed as follows:

$$\begin{aligned} ev_1([(\mathbf{a}_0, \ldots, \mathbf{a}_d, u_1, \ldots, u_{d-1})]) &:= [\mathbf{a}_0] \in CP^4, \\ ev_2([(\mathbf{a}_0, \ldots, \mathbf{a}_d, u_1, \ldots, u_{d-1})]) &:= [\mathbf{a}_d] \in CP^4. \end{aligned} \tag{5.49}$$

We also have to construct a rank $(5d + 1)$ vector bundle $\mathcal{E}_d$ on $\overline{Mp_{0,2}}(CP^4, d)$ that corresponds to the condition that the image of the rational map is contained within the quintic hypersurface. We assume that the defining equation of the quintic hypersurface is given by

$$(X_1)^5 + (X_2)^5 + \cdots + (X_5)^5. \tag{5.50}$$

Let us regard $\sum_{j=0}^{d} \mathbf{a}_j s^j t^{d-j}$ ($\mathbf{a}_0, \mathbf{a}_d \neq \mathbf{0}$) as a map φ from $\mathbf{C}^2$ to $\mathbf{C}^5$. Of course, $[(\mathbf{a}_0, \mathbf{a}_1, \ldots, \mathbf{a}_d)]$ represents a point in $Mp_{0,2}(CP^4, d)$. Then we can consider

$$(\sum_{j=0}^{d} a_j^1 s^j t^{d-j})^5 + \cdots + (\sum_{j=0}^{d} a_j^5 s^j t^{d-j})^5 = \sum_{j=0}^{5d} \varphi_j^5(\mathbf{a}_0, \ldots, \mathbf{a}_d) s^j t^{5d-j}, \tag{5.51}$$

where $\varphi_j^5(\mathbf{a}_0, \ldots, \mathbf{a}_d)$ is a homogeneous polynomial of degree 5 in a_j^i ($\mathbf{a}_j = (a_j^1, a_j^2, \ldots, a_j^N)$). If we set

$$\tilde{s}_0(\mathbf{a}_0, \ldots, \mathbf{a}_d) := (\varphi_0^5(\mathbf{a}_0, \ldots, \mathbf{a}_d), \varphi_1^5(\mathbf{a}_0, \ldots, \mathbf{a}_d), \ldots, \varphi_{5d}^5(\mathbf{a}_0, \ldots, \mathbf{a}_d)), \tag{5.52}$$

we can easily see that the image of the corresponding quasi map lies inside the hypersurface defined by (5.50) if and only if $\tilde{s}_0(\mathbf{a}_0, \ldots, \mathbf{a}_d) = \mathbf{0}$. Moreover, we can derive the following relations:

$$\begin{aligned} &\tilde{s}_0(\mu\mathbf{a}_0, \ldots, \mu\mathbf{a}_d) = (\mu^5\varphi_0^5(\mathbf{a}_0, \ldots, \mathbf{a}_d), \mu^5\varphi_1^5(\mathbf{a}_0, \ldots, \mathbf{a}_d), \ldots, \mu^5\varphi_{5d}^5(\mathbf{a}_0, \ldots, \mathbf{a}_d)), \\ &\tilde{s}_0(\mathbf{a}_0, \nu\mathbf{a}_1, \nu^2\mathbf{a}_2 \cdots, \nu^{d-1}\mathbf{a}_d) = (\varphi_0^5(\mathbf{a}_0, \ldots, \mathbf{a}_d), \nu\varphi_1^5(\mathbf{a}_0, \ldots, \mathbf{a}_d), \nu^2\varphi_2^5(\mathbf{a}_0, \ldots, \mathbf{a}_d), \ldots, \nu^{5d}\varphi_{5d}^k(\mathbf{a}_0, \ldots, \mathbf{a}_d)). \end{aligned} \tag{5.53}$$

These relations tells us that $\tilde{s}_0$ defines a section of a rank $5d + 1$ vector bundle on $Mp_{0,2}(CP^4, d)$, because we can compute transition functions of the bundle by using (5.53). Let us discuss this argument more explicitly. Since $Mp_{0,2}(CP^4, d) = (CP^4 \times \mathbf{C}^{5(d-1)} \times CP^4)/\mathbf{Z}_d$, we can take the following local coordinate system U_{ij}.

$$\begin{aligned} &\phi_{ij} : U_{ij} \subset \mathbf{C}^{5(d+1)-2} \to Mp_{0,2}(CP^4, d), \\ &\phi_{ij}(x_1, x_2, \ldots, x_4, \mathbf{y}_1, \mathbf{y}_2, \ldots \mathbf{y}_{d-1}, z_1, z_2 \cdot \cdots, z_4) = \\ &[(x_1, \ldots, x_{i-1}, 1, x_i, \ldots, x_4, \mathbf{y}_1, \ldots, \mathbf{y}_{d-1}, z_1, \ldots, z_{j-1}, 1, z_j, \ldots, z_4)], \end{aligned} \quad (5.54)$$

where $\mathbf{y}_i \in \mathbf{C}^N$. Let $(\tilde{x}_*, \tilde{\mathbf{y}}_*, \tilde{z}_*) \in U_{kl}$. We assume that $i < k$ and $j < l$ for simplicity. The coordinate transformation between U_{ij} and U_{kl} is given by

$$\begin{aligned} &x_m = \frac{\tilde{x}_m}{\tilde{x}_i} \;\; (m \leq i-1), \;\; x_m = \frac{\tilde{x}_{m+1}}{\tilde{x}_i} \;\; (i \leq m \leq k-2), \\ &x_{k-1} = \frac{1}{\tilde{x}_i}, \;\; x_m = \frac{\tilde{x}_m}{\tilde{x}_i} \;\; (k \leq m \leq 4), \\ &z_m = \frac{\tilde{z}_m}{\tilde{z}_j} \;\; (m \leq j-1), \;\; z_m = \frac{\tilde{z}_{m+1}}{\tilde{z}_j} \;\; (j \leq m \leq l-2), \\ &z_{l-1} = \frac{1}{\tilde{z}_j}, \;\; z_m = \frac{\tilde{z}_m}{\tilde{z}_j} \;\; (l \leq m \leq 4), \\ &\mathbf{y}_m = \frac{1}{(\tilde{x}_i)^{\frac{d-m}{d}}(\tilde{z}_j)^{\frac{m}{d}}}\tilde{\mathbf{y}}_m. \end{aligned} \quad (5.55)$$

If we represent the section s_0 on U_{ij} by

$$\tilde{s}_0(\phi_{ij}(x_*, \mathbf{y}_*, z_*)) = (\varphi_0(x_*, \mathbf{y}_*, z_*), \varphi_1(x_*, \mathbf{y}_*, z_*), \ldots, \varphi_{5d}(x_*, \mathbf{y}_*, z_*)),$$

we obtain the following relation:

$$\frac{1}{(\tilde{x}_i)^{\frac{5d-m}{d}}(\tilde{z}_j)^{\frac{m}{d}}}\varphi_m(\tilde{x}_*, \tilde{\mathbf{y}}_*, \tilde{z}_*) = \varphi_m(x_*, \mathbf{y}_*, z_*), \;\; (m = 0, 1, 2, \ldots, kd). \quad (5.56)$$

Therefore, we can regard $\tilde{s}_0$ as a section of the rank $5d + 1$ bundle whose transition function is given by

$$(\tilde{x}_i)^{\frac{5d-m}{d}}(\tilde{z}_j)^{\frac{m}{d}} e_m = e_m, \;\; (m = 0, 1, 2, \ldots, 5d), \quad (5.57)$$

where e_m (resp. e_m) is the base of trivialization on U_{ij} (resp. U_{kl}). We denote this vector bundle on $Mp_{0,2}(CP^4, d)$ by $\mathscr{E}_d$. From (5.57), we can see that

$\mathscr{E}_d \simeq \oplus_{m=0}^{5d}\big(\mathscr{O}_{CP^4}(\frac{5d-m}{d}) \otimes \mathscr{O}_{CP^4}(\frac{m}{d})\big)$ as a vector bundle on $Mp_{0,2}(CP^4, d)$. Next, we extend $\mathscr{E}_d$ to $\overline{Mp}_{0,2}(CP^4, d)$. Let us consider the locus in $\overline{Mp}_{0,2}(CP^4, d)$, where

$$\begin{aligned} &u_{d_j} = 0, \ \ (1 \leq d_1 < d_2 < \cdots < d_{l-1} \leq d-1), \\ &u_j \neq 0, \ \ (j \notin \{d_1, d_2, \ldots, d_{l-1}\}). \end{aligned} \tag{5.58}$$

We denote this locus by $U_{(d_0,d_1,\ldots,d_l)}$ $(d_0 := 0,\ d_l := d)$. As was discussed in the previous section, $U_{(d_0,d_1,\ldots,d_l)}$ is identified with

$$Mp_{0,2}(CP^4, d_1 - d_0) \underset{CP^4}{\times} Mp_{0,2}(CP^4, d_2 - d_1) \underset{CP^4}{\times} \cdots \underset{CP^4}{\times} Mp_{0,2}(CP^4, d_l - d_{l-1}), \tag{5.59}$$

and its point is represented by a chain of quasi maps:

$$\bigcup_{j=1}^{l}\Big(\sum_{h=0}^{d_j-d_{j-1}} \mathbf{a}_{d_{j-1}+h}(s_j)^h(t_j)^{d_j-d_{j-1}-h}\Big). \tag{5.60}$$

For each $Mp_{0,2}(CP^4, d_j - d_{j-1})$, we have the $(5(d_j - d_{j-1}) + 1)$-dimensional vector bundle $\mathscr{E}_{d_j-d_{j-1}}$. We then introduce a map $p_j : U_{(d_0,d_1,\ldots,d_l)} \to CP^4$ $(j = 1, 2, \ldots, l-1)$ defined by

$$p_j\Big(\bigcup_{j=1}^{l}\Big(\sum_{h=0}^{d_j-d_{j-1}} \mathbf{a}_{d_{j-1}+h}(s_j)^h(t_j)^{d_j-d_{j-1}-h}\Big)\Big) = [\mathbf{a}_{d_j}] \in CP^4. \tag{5.61}$$

With this set-up, we define $\mathscr{E}_d|_{U_{(d_0,d_1,\ldots,d_l)}}$ by the following exact sequence:

$$0 \to \mathscr{E}_d|_{U_{(d_0,d_1,\ldots,d_l)}} \to \bigoplus_{j=1}^{l} \mathscr{E}_{d_j-d_{j-1}} \to \bigoplus_{j=1}^{l-1} p_j^*\mathscr{O}_{CP^4}(5) \to 0. \tag{5.62}$$

$\mathscr{E}_d|_{U_{(d_0,d_1,\ldots,d_l)}}$ also has rank $5d + 1$. In this way, we extend $\mathscr{E}_d$ to whole $\overline{Mp}_{0,2}(CP^4, d)$.

5.4 Fixed Point Theorem and Computation of $w(\mathscr{O}_{h^a}\mathscr{O}_{h^b})_{0,d}$

5.4.1 *Fixed Point Theorem*

In this subsection, we introduce the fixed point theorem proved by Bott in [5]. This enables us to compute the intersection number $w(\mathscr{O}_{h^a}\mathscr{O}_{h^b})_{0,d}$ defined in the previous section explicitly. We only explain the assertion of the theorem. Let E_i, $(i = 1, \ldots, n)$ be vector bundles on a complex space X. The rank of the vector bundle E_i is given by r_i. What we want to evaluate is the following intersection number,

$$\int_X \prod_{i=1}^{n} c_{m_i}(E_i), \qquad (\sum_{i=1}^{n} m_i = \dim_{\mathbf{C}}(X)), \tag{5.63}$$

where $c_{m_i}(E_i)$ is the m_i-th Chern class of the vector bundle E_i. Now, we assume that the multiplicative group $\mathbf{C}^{\times}$ of complex numbers acts on X via the following map,

$$\varphi : \mathbf{C}^{\times} \times X \to X. \tag{5.64}$$

Intuitively, this action corresponds to a "flow" on X induced from some complex tangent vector field on X. With this set-up, we consider the fixed point set of this flow, i.e., the set of points $x \in X$ that satisfy $\varphi(e^u, x) = x$ for arbitrary $e^u \in \mathbf{C}^{\times}$. Let F_j $(j = 1, 2, \ldots, N)$ be connected components of the fixed point set. The fixed point theorem gives us the formula that represents the integral in (5.63) as the sum of integrals on the connected components F_j $(j = 1, 2, \ldots, N)$.

Let us introduce differential forms to integrate on the connected component F_j. First, we decompose the restriction $E_i|_{F_j}$ of the vector bundle E_i to F_j into a direct sum with respect to the eigenvalue of the $\mathbf{C}^{\times}$ action:

$$E_i|_{F_j} = \bigoplus_{m=1}^{k_{i,j}} \mathcal{E}_{i,j}^{(m)}. \tag{5.65}$$

Let $r_{i,j}^{(m)}$ be the rank of $\mathcal{E}_{i,j}^{(m)}$. Then we have the relation $\sum_{m=1}^{k_{i,j}} r_{i,j}^{(m)} = r_i$. Here, $\mathcal{E}_{i,j}^{(m)}$ responds in the following way:

$$\mathcal{E}_{i,j}^{(m)} \to e^{v_{i,j}^{(m)} u} \mathcal{E}_{i,j}^{(m)}, \tag{5.66}$$

under the action of $e^u \in \mathbf{C}^{\times}$, and we call the above $v_{i,j}^{(m)}$ the eigenvalue of $\mathcal{E}_{i,j}^{(m)}$ under the $\mathbf{C}^{\times}$ action. We further introduce the virtual decomposition $\mathcal{E}_{i,j}^{(m)} = \oplus_{h=1}^{r_{i,j}^{(m)}} e_{i,j}^{(m,h)}$ of $\mathcal{E}_{i,j}^{(m)}$ into line bundles and define the formal first Chern class $c_1(e_{i,j}^{(m,h)})$ of the line bundle $e_{i,j}^{(m,h)}$ by the formula

$$c(\mathcal{E}_{i,j}^{(m)}) = \sum_{n=0}^{r_{i,j}^{(m)}} t^n c_n(\mathcal{E}_{i,j}^{(m)}) = \prod_{h=1}^{r_{i,j}^{(m)}} (1 + tc_1(e_{i,j}^{(m,h)})). \tag{5.67}$$

Here, we assume that we can determine $c(\mathcal{E}_{i,j}^{(m)})$ explicitly. With this set-up, we introduce the n-th equivariant Chern class $\tilde{c}_n(E_i|_{F_j})$ of $E_i|_{F_j}$ defined as follows:

$$\sum_{n=0}^{r_i} t^n \tilde{c}_i(E_i|_{F_j}) := \prod_{m=1}^{k_{i,j}} \prod_{h=1}^{r_{i,j}^{(m)}} (1 + t(v_{i,j}^{(m)} + c_1(e_{i,j}^{(m,h)}))). \tag{5.68}$$

As can be seen from the above definition, $\tilde{c}_n(E_i|_{F_j})$ is usually given as a linear combination of differential forms of degree $2j$ ($j = 0, 1, \ldots, n$). In some cases, it turns out to be just a number (0 form). The definitions we have introduced so far are quite complicated. Therefore, we explain later how to use these definitions by using rather simple but non-trivial examples.

In order to describe differential forms to integrate on F_j, we have to consider the holomorphic normal bundle N_j of F_j in X, in addition to equivariant Chern classes of $E_i|_{F_j}$. In the same way as the case of $E_i|_{F_j}$, we first decompose N_j into a direct sum with respect to the eigenvalue under the $\mathbf{C}^{\times}$ action:

$$N_j = \bigoplus_{m=1}^{l_j} \mathcal{N}_j^{(m)}. \tag{5.69}$$

Let s_j be the rank of N_j and $s_j^{(m)}$ be the rank of $\mathcal{N}_j^{(m)}$. Then the relation $\sum_{j=1}^{l_j} s_j^{(m)} = s_j$ holds. We denote the eigenvalue of $\mathcal{N}_j^{(m)}$ under the $\mathbf{C}^{\times}$ action by $\kappa_j^{(m)}$. We further introduce the virtual decomposition $\mathcal{N}_j^{(m)} = \oplus_{h=1}^{s_j^{(m)}} n_j^{(m,h)}$ of $\mathcal{N}_j^{(m)}$ into line bundles and define the formal first Chern class $c_1(n_j^{(m,h)})$ by the following formula:

$$c(\mathcal{N}_j^{(m)}) = \sum_{n=0}^{s_j^{(m)}} t^n c_n(\mathcal{N}_j^{(m)}) = \prod_{h=1}^{s_j^{(m)}} (1 + tc_1(n_j^{(m,h)})). \tag{5.70}$$

With this set-up, we define the s_j-th equivariant Chern class $\tilde{c}_{s_j}(N_j)$ of N_j by the following formula:

$$\tilde{c}_{s_j}(N_j) := \prod_{m=1}^{l_j} \prod_{h=1}^{s_j^{(m)}} (\kappa_j^{(m)} + c_1(n_j^{(m,h)})). \tag{5.71}$$

From this formula, we can see that $\tilde{c}_{s_j}(N_j)$ is also given by a linear combination of differential forms whose degrees are even and no more than $2s_j$ in the same manner as the equivariant Chern classes of $E_i|_{F_j}$.

With the aid of the rather complicated definitions we have introduced so far, we can state the assertion of the fixed point theorem, which we will use to test the mirror symmetry hypothesis.

Theorem 5.4.1

$$\int_X \prod_{i=1}^{n} c_{m_i}(E_i) = \sum_{j=1}^{N} \int_{F_j} \frac{\prod_{i=1}^{n} \tilde{c}_{m_i}(E_i|_{F_j})}{\tilde{c}_{s_j}(N_j)}. \tag{5.72}$$

In the r.h.s. of the above equation, there appears "division" by an equivariant Chern class, which seems curious at first sight. As can be seen from the definition of an equivariant Chern class, it always contains degree 0 constant term, and the division operation can be done by using a formal power series expansion with respect to Chern classes of positive degree.

5.4.2 Computation of $w(\mathcal{O}_{h^a}\mathcal{O}_{h^b})_{0,d}$

In this subsection, we compute the intersection number $w(\mathcal{O}_{h^a}\mathcal{O}_{h^b})_{0,d}$ by using the fixed point theorem. For this purpose, we introduce the following $\mathbf{C}^{\times}$ action on $\overline{Mp}_{0,2}(CP^4,d)$:

$$[(e^{\lambda_0 t}\mathbf{a}_0, e^{\lambda_1 t}\mathbf{a}_1, \ldots, e^{\lambda_{d-1} t}\mathbf{a}_{d-1}, e^{\lambda_d t}\mathbf{a}_d, u_1, u_2, \ldots, u_{d-1})]. \quad (5.73)$$

The fixed point sets of $\overline{Mp}_{0,2}(N,d)$ consist of connected components, each of which come from $U_{(d_0,d_1,\ldots,d_l)}$ defined in the previous section. We denote the connected component that comes from $U_{(d_0,d_1,\ldots,d_l)}$ by $F_{(d_0,d_1,\ldots,d_l)}$. Explicitly, a point in $F_{(d_0,d_1,\ldots,d_l)}$ is represented by the following chain of polynomial maps:

$$\bigcup_{j=1}^{l}(\mathbf{a}_{d_{j-1}}(s_j)^{d_j-d_{j-1}} + \mathbf{a}_{d_j}(t_j)^{d_j-d_{j-1}}). \quad (5.74)$$

Note here that $(\mathbf{a}_{d_{j-1}}(s_j)^{d_j-d_{j-1}} + \mathbf{a}_{d_j}(t_j)^{d_j-d_{j-1}})$ is the $\mathbf{Z}_{d_j-d_{j-1}}$ singularity in $Mp_{0,2}(CP^4, d_j - d_{j-1})$. We can easily see from (5.74) that $F_{(d_0,d_1,\ldots,d_l)}$ is set-theoretically isomorphic to $\prod_{j=0}^{l}(CP^4)_{d_j}$, where $(CP^4)_{d_j}$ is the CP^4 whose point is given by $[\mathbf{a}_{d_j}]$.

Let us consider the contribution to $w(\mathcal{O}_{h^a}\mathcal{O}_{h^b})_{0,d}$ from $F_{(d_0,d_1,\ldots,d_l)}$. We start from the case of $F_{(0,d)} \subset U_{(0,d)} = Mp_{0,2}(CP^4,d)$. First, we have to determine the normal bundle of $F_{(0,d)}$ in $Mp_{0,2}(CP^4,d)$. We already know from the previous discussion that

$$Mp_{0,2}(CP^4,d) = \{([\mathbf{a}_0], \mathbf{y}_1, \ldots, \mathbf{y}_{d-1}, [\mathbf{a}_d]) \mid [\mathbf{a}_0], [\mathbf{a}_d] \in CP^4,\ \mathbf{y}_i \in C^5\}/\mathbf{Z}_d. \quad (5.75)$$

Therefore, the normal bundle is given by $\displaystyle\bigoplus_{i=1}^{d-1}\bigoplus_{j=1}^{5}\frac{\partial}{\partial y_i^j}$. From the discussion of the previous section, we can see that $\frac{\partial}{\partial y_i^j}$ is isomorphic to $\mathcal{O}_{(CP^4)_0}(\frac{d-i}{d}) \otimes \mathcal{O}_{(CP^4)_d}(\frac{i}{d})$ as a vector bundle on $F_{(0,d)}$ and its first Chern class is given by

$$\frac{d-i}{d}h_0 + \frac{i}{d}h_d, \quad (5.76)$$

where h_{d_i} is the hyperplane class of $(CP^4)_{d_i}$. On the other hand, the flow in (5.73) acts on y_i^j as $y_i^j \to e^{\left(\lambda_i-(\frac{d-i}{d}\lambda_0+\frac{i}{d}\lambda_d)\right)t}y_i^j$, and the character of the flow on $\frac{\partial}{\partial y_i^j}$ is given by

$$\frac{d-i}{d}\lambda_0+\frac{i}{d}\lambda_d-\lambda_i. \tag{5.77}$$

Next, we consider the equivariant top Chern class of $\mathcal{E}_d$ on $F_{(0,d)}$. Since $\mathcal{E}_d$ is identified with $\oplus_{m=0}^{5d}\left(\mathcal{O}_{(CP^4)_0}(\frac{kd-m}{d})\otimes\mathcal{O}_{(CP^4)_d}(\frac{m}{d})\right)$ as a vector bundle on $Mp_{0,2}(CP^4,d)$, its equivariant top Chern class on $F_{(0,d)}$ is given by

$$\prod_{m=0}^{5d}\left(\frac{(5d-m)(h_0+\lambda_0)+m(h_d+\lambda_d)}{d}\right). \tag{5.78}$$

From the definition of the evaluation map for $Mp_{0,2}(CP^4,d)$ in (5.49), we can easily see that the equivariant representation of $ev_1^*(h^a)$ (resp. $ev_2^*(h^b)$) on $F_{(0,d)}$ is given by $(h_0+\lambda_0)^a$ (resp. $(h_d+\lambda_d)^b$). Finally, we have to remember that $F_{(0,d)}$ is also the singular locus on which $\mathbf{Z}_d$ acts. Therefore, we have to divide the results of integration on $F_{(0,d)}$ by d. Putting these results altogether, the contribution from $F_{(0,d)}$ becomes,

$$\frac{1}{d}\int_{(CP^4)_0}\int_{(CP^4)_d}(h_0+\lambda_0)^a\frac{\prod_{m=0}^{5d}\left(\frac{(5d-m)(h_0+\lambda_0)+m(h_d+\lambda_d)}{d}\right)}{\prod_{i=1}^{d-1}\left(\frac{(d-i)(h_0+\lambda_0)+i(h_d+\lambda_d)}{d}-\lambda_i\right)^5}(h_d+\lambda_d)^b. \tag{5.79}$$

We then consider the contribution from $F_{(d_0,d_1,\ldots,d_l)}$ $(l\geq 2)$. As for the normal bundle, we have additional factors coming from "smoothing the nodal singularities" of the image of the chain of quasi maps, which are given by $[\mathbf{a}_{d_j}]$ ($j=1,2,\ldots,l-1$). This factor is identified with the vector bundle $\frac{d}{d(\frac{s_j}{t_j})}\otimes\frac{d}{d(\frac{t_{j+1}}{s_{j+1}})}$ and its equivariant first Chern class is given by

$$\frac{h_{d_j}+\lambda_{d_j}-h_{d_{j-1}}-\lambda_{d_{j-1}}}{d_j-d_{j-1}}+\frac{h_{d_j}+\lambda_{d_j}-h_{d_{j+1}}-\lambda_{d_{j+1}}}{d_{j+1}-d_j}. \tag{5.80}$$

The equivariant top Chern class of $\mathcal{E}_d$ on $F_{(d_0,d_1,\ldots,d_l)}$ can be read off from the exact sequence in (5.62) as follows:

$$\frac{\prod_{j=1}^{l}\prod_{m=0}^{5(d_j-d_{j-1})}\left(\frac{(5(d_j-d_{j-1})-m)(h_{d_{j-1}}+\lambda_{d_{j-1}})+m(h_{d_j}+\lambda_{d_j})}{d_j-d_{j-1}}\right)}{\prod_{j=1}^{l-1}5(h_{d_j}+\lambda_{d_j})}. \tag{5.81}$$

Combining these additional factors with the consideration in the case of $F_{(0,d)}$, we can write down the contribution that comes from $F_{(d_0,d_1,\cdots,d_l)}$:

$$\frac{1}{\prod_{j=1}^{l}(d_j-d_{j-1})}\int_{(CP^4)_{d_0}}\cdots\int_{(CP^4)_{d_l}}(h_{d_0}+\lambda_{d_0})^a\times$$
$$\frac{\prod_{j=1}^{l}\prod_{m=0}^{5(d_j-d_{j-1})}\left(\frac{(5(d_j-d_{j-1})-m)(h_{d_{j-1}}+\lambda_{d_{j-1}})+m(h_{d_j}+\lambda_{d_j})}{d_j-d_{j-1}}\right)}{\prod_{j=1}^{l}\prod_{i=1}^{d_j-d_{j-1}-1}\left(\frac{(d_j-d_{j-1}-i)(h_{d_{j-1}}+\lambda_{d_{j-1}})+i(h_{d_j}+\lambda_{d_j})}{d_j-d_{j-1}}-\lambda_{d_{j-1}+i}\right)^5}\times$$
$$\frac{1}{\prod_{j=1}^{l-1}\left(\frac{h_{d_j}+\lambda_{d_j}-h_{d_{j-1}}-\lambda_{d_{j-1}}}{d_j-d_{j-1}}+\frac{h_{d_j}+\lambda_{d_j}-h_{d_{j+1}}-\lambda_{d_{j+1}}}{d_{j+1}-d_j}\right)(5(h_{d_j}+\lambda_{d_j}))}(h_{d_l}+\lambda_{d_l})^b.$$
(5.82)

Let $\frac{1}{2\pi\sqrt{-1}}\oint_{C_{(0)}}dz$ be operation of taking residue at $z=0$. Since $\int_{CP^4}h^a=\frac{1}{2\pi\sqrt{-1}}\oint_{C_{(0)}}\frac{dz}{z^5}z^a$, we obtain the following closed formula for $w(\mathcal{O}_{h^a}\mathcal{O}_{h^b})_{0,d}$:

$$w(\mathcal{O}_{h^a}\mathcal{O}_{h^b})_{0,d}=$$
$$\sum_{0=d_0<d_1<\cdots<d_{l-1}<d_l=d}\frac{1}{\prod_{j=1}^{l}(d_j-d_{j-1})}\frac{1}{(2\pi\sqrt{-1})^{l+1}}\oint_{C_{(0)}}\frac{dz_{d_0}}{(z_{d_0})^5}\cdots\oint_{C_{(0)}}\frac{dz_{d_l}}{(z_{d_l})^5}\times$$
$$(z_{d_0}+\lambda_{d_0})^a\frac{\prod_{j=1}^{l}\prod_{m=0}^{5(d_j-d_{j-1})}\left(\frac{(5(d_j-d_{j-1})-m)(z_{d_{j-1}}+\lambda_{d_{j-1}})+m(z_{d_j}+\lambda_{d_j})}{d_j-d_{j-1}}\right)}{\prod_{j=1}^{l}\prod_{i=1}^{d_j-d_{j-1}-1}\left(\frac{(d_j-d_{j-1}-i)(z_{d_{j-1}}+\lambda_{d_{j-1}})+i(z_{d_j}+\lambda_{d_j})}{d_j-d_{j-1}}-\lambda_{d_{j-1}+i}\right)^5}\times$$
$$\frac{1}{\prod_{j=1}^{l-1}\left(\frac{z_{d_j}+\lambda_{d_j}-z_{d_{j-1}}-\lambda_{d_{j-1}}}{d_j-d_{j-1}}+\frac{z_{d_j}+\lambda_{d_j}-z_{d_{j+1}}-\lambda_{d_{j+1}}}{d_{j+1}-d_j}\right)(5(z_{d_j}+\lambda_{d_j}))}(z_{d_l}+\lambda_{d_l})^b. \quad (5.83)$$

In the above formula, we can integrate the variables z_{d_j} in an arbitrary order. The formula (5.83) has the form of a residue integral and we can take the non-equivariant limit $\lambda_j\to 0$. This operation makes the formula simpler. For simplicity, we introduce the following notations. We define the following two polynomials in z and w:

$$e(d;z,w):=\prod_{j=0}^{5d}\left(\frac{jz+(5d-j)w}{d}\right)$$
$$t(d;z,w):=\prod_{j=1}^{d-1}\left(\frac{jz+(d-j)w}{d}\right)^5. \quad (5.84)$$

We also introduce the ordered partition of a positive integer d:

Definition 5.4.1 Let OP_d be the set of ordered partitions of a positive integer d:

$$OP_d=\{\sigma_d=(d_1,d_2,\ldots,d_{l(\sigma_d)})\mid\sum_{j=1}^{l(\sigma_d)}d_j=d\ ,\ d_j\in\mathbf{N}\}. \quad (5.85)$$

In (5.85), we denoted the length of the ordered partition σ_d by $l(\sigma_d)$.

The increasing sequence of integers $(d_0, d_1, \ldots, d_l)$ $(0 = d_0 < d_1 < \cdots < d_{l-1} < d_l = d)$ used in (5.83) can be replaced by the ordered partition $\sigma_d = (\tilde{d}_1, \tilde{d}_2, \ldots, \tilde{d}_l) \in OP_d$ if we use the following correspondence:

$$\tilde{d}_j = d_j - d_{j-1}, \quad (j = 1, 2, \ldots, l). \tag{5.86}$$

With this set-up, we can simplify the formula for $w(\mathcal{O}_{h^a}\mathcal{O}_{h^b})_{0,d}$ after taking the non-equivariant limit, by relabeling the subscript of $z_*'s$ as follows:

$$w(\mathcal{O}_{h^a}\mathcal{O}_{h^b})_{0,d} = \sum_{\sigma_d \in OP_d} \frac{1}{(2\pi\sqrt{-1})^{l(\sigma_d)+1}\prod_{j=0}^{l(\sigma_d)} d_j} \oint_{C_0} \frac{dz_0}{(z_0)^5} \cdots \oint_{C_0} \frac{dz_{l(\sigma_d)}}{(z_{l(\sigma_d)})^5} (z_0)^a \times$$
$$\times \prod_{j=1}^{l(\sigma_d)-1} \frac{1}{\left(\frac{z_j - z_{j-1}}{d_j} + \frac{z_j - z_{j+1}}{d_{j+1}}\right) k z_j} \prod_{j=1}^{l(\sigma_d)} \frac{e(d_j; z_{j-1}, z_j)}{t(d_j; z_{j-1}, z_j)} (z_{l(\sigma_d)})^b. \tag{5.87}$$

Remark 5.4.1 After taking the non-equivariant limit, we have to take care of the order of integration of z_j's. In (5.87), we have to integrate $z_j's$ in all the summands of the formula in descending (or ascending) order of the subscript j.

We can further simplify the formula (5.87) in the following form.

Theorem 5.4.2 *Let* $\frac{1}{2\pi\sqrt{-1}}\oint_{C_{(0,\frac{z_{j-1}+z_{j+1}}{2})}} dz_j$ *be the operation of taking residues at* $z_j = 0$ *and* $z_j = \frac{z_{j-1}+z_{j+1}}{2}$. *Then the following equality holds:*

$$w(\mathcal{O}_{h^a}\mathcal{O}_{h^b})_{0,d} = \frac{1}{(2\pi\sqrt{-1})^{d+1}} \oint_{C_0} \frac{dz_0}{(z_0)^5} \oint_{C_1} \frac{dz_1}{(z_1)^5} \cdots \oint_{C_{d-1}} \frac{dz_{d-1}}{(z_{d-1})^5} \oint_{C_d} \frac{dz_d}{(z_d)^5} \times$$
$$(z_0)^a \frac{\prod_{j=1}^{d} e(1; z_{j-1}, z_j)}{\prod_{i=1}^{d-1} 5z_i(2z_i - z_{i-1} - z_{i+1}))} (z_d)^b, \tag{5.88}$$

where $\frac{1}{2\pi\sqrt{-1}}\oint_{C_j} dz_j$ $(j = 1, \ldots, d-1)$ *represents* $\frac{1}{2\pi\sqrt{-1}}\oint_{C_{(0,\frac{z_{j-1}+z_{j+1}}{2})}} dz_j$ *and* $\frac{1}{2\pi\sqrt{-1}}\oint_{C_j} dz_j$ $(j = 0, d)$ *represents* $\frac{1}{2\pi\sqrt{-1}}\oint_{C_{(0)}} dz_j$.

Proof We first pay attention to the fact that $\oint_{C_j} dz_j$ is decomposed into $\oint_{C_0} dz_j + \oint_{C_{\frac{z_{j-1}+z_{j+1}}{2}}} dz_j$ for $j = 1, 2, \ldots, d-1$. Therefore, the r.h.s. of (5.88) can be rewritten as follows:

$$\frac{1}{(2\pi\sqrt{-1})^{d+1}}\sum_{n=0}^{d-1}\sum_{1\leq j_1<j_2<\cdots<j_n\leq d-1}\oint_{C_0}dz_0\cdots\oint_{C_0}dz_{j_1-1}\oint_{C_{\frac{z_{j_1-1}+z_{j_1+1}}{2}}}dz_{j_1}\times$$
$$\oint_{C_0}dz_{j_1+1}\cdots\oint_{C_0}dz_{j_2-1}\oint_{C_{\frac{z_{j_2-1}+z_{j_2+1}}{2}}}dz_{j_2}\oint_{C_0}dz_{j_2+1}\cdots\cdots\oint_{C_0}dz_{j_n-1}$$
$$\oint_{C_{\frac{z_{j_n-1}+z_{j_n+1}}{2}}}dz_{j_n}\oint_{C_0}dz_{j_n+1}\cdots\oint_{C_0}dz_d\frac{1}{\prod_{i=0}^{d}(z_i)^5}\times(z_0)^a\frac{\prod_{j=1}^{d}e(1;z_{j-1},z_j)}{\prod_{i=1}^{d-1}5z_i(2z_i-z_{i-1}-z_{i+1})}(z_d)^b. \tag{5.89}$$

Then we change integration variables of the summand that corresponds to $1\leq j_1<j_2<\cdots<j_n\leq d-1$ as follows:

$$\begin{aligned} u_i&=z_i \quad \text{if } i\notin\{j_1,j_2,\ldots,j_n\},\\ u_i&=2z_i-z_{i-1}-z_{i+1} \quad \text{if } i\in\{j_1,j_2,\ldots,j_n\}. \end{aligned} \tag{5.90}$$

Let $\{i_1,i_2,\ldots,i_{l-1}\}$ be $\{1,2,\ldots,d-1\}-\{j_1,j_2,\ldots,j_n\}$, where

$$0=:i_0<i_1<i_2<\cdots<i_{l-1}<i_l:=d,\ \ l=d-n. \tag{5.91}$$

Inversion of (5.90) results in

$$\begin{aligned} z_j(u_*)&=u_j \quad \text{if } j\in\{i_0,i_1,\ldots,i_l\},\\ z_j(u_*)&=\frac{(i_m-j)u_{i_{m-1}}+(j-i_{m-1})u_{i_m}+\sum_{h=i_{m-1}+1}^{i_m-1}C_h^ju_h}{i_m-i_{m-1}} \quad \text{if } i_{m-1}+1\leq j\leq i_m-1, \end{aligned} \tag{5.92}$$

where C_h^j is some positive integer. The Jacobian of this coordinate change is given by

$$\frac{1}{\prod_{m=1}^{l}(i_m-i_{m-1})}. \tag{5.93}$$

In this way, the term corresponding to $1\leq j_1<j_2<\cdots<j_n\leq d-1$ in (5.89) can be rewritten as follows:

$$\frac{1}{(2\pi\sqrt{-1})^{d+1}}\frac{1}{\prod_{m=1}^{l}(i_m-i_{m-1})}\oint_{C_0}du_0\oint_{C_0}du_1\cdots\oint_{C_0}du_d\times$$
$$\frac{(z_0(u_*))^a(z_d(u_*))^b}{(z_0(u_*))^5(z_d(u_*))^5}\times\frac{1}{\prod_{m=1}^{l-1}\left((z_{i_m}(u_*))^5(2z_{i_m}(u_*)-z_{i_m-1}(u_*)-z_{i_m+1}(u_*))\right)}\times$$
$$\frac{1}{\prod_{m=1}^{l}\prod_{j=i_{m-1}+1}^{i_m-1}u_j\cdot(z_j(u_*))^5}\times\prod_{m=1}^{l-1}\frac{1}{5z_{i_m}(u_*)}\prod_{m=1}^{l}\frac{\prod_{j=i_{m-1}+1}^{i_m}e(1;z_{j-1}(u_*),z_j(u_*))}{\prod_{j=i_{m-1}+1}^{i_m-1}5z_j(u_*)}. \tag{5.94}$$

Looking at (5.94), we observe that the integrand has only a simple pole at $u_{j_h} = 0$ $(h = 1, 2, \ldots, n)$. Therefore, we can take the residue of u_{j_h} before u_{i_m} $(m = 0, 1, \ldots, l)$. After this operation, (5.92) reduces to

$$z_j(u_*) = u_j \quad \text{if } j \in \{i_0, i_1, \cdots, i_l\},$$
$$z_j(u_*) = \frac{(i_m - j)u_{i_{m-1}} + (j - i_{m-1})u_{i_m}}{i_m - i_{m-1}} \quad \text{if } i_{m-1} + 1 \leq j \leq i_m - 1. \tag{5.95}$$

With (5.95) and some algebra, we can easily derive

$$2z_{i_m}(u_*) - z_{i_m-1}(u_*) - z_{i_m+1}(u_*) = \frac{u_{i_m} - u_{i_{m-1}}}{i_m - i_{m-1}} + \frac{u_{i_m} - u_{i_{m+1}}}{i_{m+1} - i_m},$$

$$\prod_{m=1}^{l} \prod_{j=i_{m-1}+1}^{i_m-1} (z_j(u_*))^5 = \prod_{m=1}^{l} t(i_m - i_{m-1}; u_{i_{m-1}}, u_{i_m}),$$

$$\frac{\prod_{j=i_{m-1}+1}^{i_m} e(1; z_{j-1}(u_*), z_j(u_*))}{\prod_{j=i_{m-1}+1}^{i_m-1} 5z_j(u_*)} = e(i_m - i_{m-1}; u_{i_{m-1}}, u_{i_m}).$$

and (5.94) equals

$$\frac{1}{(2\pi\sqrt{-1})^{l+1}} \cdot \frac{1}{\prod_{m=1}^{l}(i_m - i_{m-1})} \oint_{C_0} du_0 \oint_{C_0} du_{i_1} \oint_{C_0} du_{i_2} \cdots \oint_{C_0} du_{i_l} \times$$
$$\frac{u_0^a u_d^b}{(u_0)^5 (u_d)^5} \cdot \frac{1}{\prod_{m=1}^{l-1}((u_{i_m})^5(\frac{u_{i_m}-u_{i_{m-1}}}{i_m-i_{m-1}} + \frac{u_{i_m}-u_{i_{m+1}}}{i_{m+1}-i_m}))5u_{i_m}} \times \prod_{m=1}^{l} \frac{e(i_m - i_{m-1}; u_{i_{m-1}}, u_{i_m})}{t(i_m - i_{m-1}; u_{i_{m-1}}, u_{i_m})}. \tag{5.96}$$

By setting $d_m = i_m - i_{m-1}$ and $z_m = u_{i_m}$, (5.96) turns out be the summand of the l.h.s. of (5.88) corresponding to $\sigma_d = (d_1, d_2, \ldots, d_l) \in OP_d$ $(l = l(\sigma_d))$.

Remark 5.4.2 Though we omit the detailed proof, we can show that the r.h.s. of (5.88) does not depend on the order of integration of variables z_j.

5.5 Reconstruction of Mirror Symmetry Computation

From the dimensional counting, $w(\mathcal{O}_{h^a}\mathcal{O}_{h^b})_{0,d}$ is non-zero only if $a + b = 2$. By using Theorem 5.4.2, we can obtain numerical results of non-vanishing $w(\mathcal{O}_{h^a}\mathcal{O}_{h^b})_{0,d}$'s for low degrees:

$$w(\mathcal{O}_{h^0}\mathcal{O}_{h^2})_{0,1} = 3850,\ w(\mathcal{O}_{h^0}\mathcal{O}_{h^2})_{0,2} = 3589125,\ w(\mathcal{O}_{h^0}\mathcal{O}_{h^2})_{0,3} = \frac{16126540000}{3},$$
$$w(\mathcal{O}_{h^1}\mathcal{O}_{h^1})_{0,1} = 6725,\ w(\mathcal{O}_{h^1}\mathcal{O}_{h^1})_{0,2} = \frac{16482625}{2},\ w(\mathcal{O}_{h^1}\mathcal{O}_{h^1})_{0,3} = \frac{44704818125}{3}. \quad (5.97)$$

Let us consider here the generating function:

$$t(u) := x + \sum_{d=1}^{\infty} \frac{w(\mathcal{O}_{h^0}\mathcal{O}_{h^2})_{0,d}}{5} e^{du} = u + 770e^{u} + 717825e^{2u} + \frac{3225308000}{3}e^{3u} + \cdots. \quad (5.98)$$

This is nothing but the mirror map given in Chap. 4! If we introduce another generating function:

$$F(u) := 5u + \sum_{d=1}^{\infty} w(\mathcal{O}_h\mathcal{O}_h)_{0,d}e^{du} = 5x + 6725e^{u} + \frac{16482625}{2}e^{2u} + \frac{44704818125}{3}e^{3u} + \cdots, \quad (5.99)$$

$F(u(t))$ gives

$$F(u(t)) = 5t + 2875e^{t} + \frac{4876875}{2}e^{2t} + \frac{8564575000}{3}e^{3t} + \cdots, \quad (5.100)$$

which reproduces the instanton correction predicted by the mirror symmetry hypothesis!

Motivated by these numerical results, we reconstruct the B-model computation of a quintic hypersurface from the intersection number $w(\mathcal{O}_{h^a}\mathcal{O}_{h^b})_{0,d}$. For brevity, we introduce the notation:

Definition 5.5.1

$$e(x, y) := e(1; x, y) = \prod_{i=0}^{5}(ix + (5-i)y). \quad (5.101)$$

Next, define a formal intersection number $w(\mathcal{O}_{h^3}\mathcal{O}_{h^{-1}})_{0,d}$ by

$$w(\mathcal{O}_{h^3}\mathcal{O}_{h^{-1}})_{0,d} := \prod_{j=0}^{d}\left(\frac{1}{2\pi\sqrt{-1}}\oint_{C_j}\frac{dz_j}{(z_j)^5}\right)(z_0)^3\left(\prod_{j=1}^{d}e(z_{j-1}, z_j)\right)\left(\prod_{j=1}^{d-1}\frac{1}{5z_j(2z_j - z_{j-1} - z_{j+1})}\right)\frac{1}{z_d}. \quad (5.102)$$

Then we introduce the following generating functions:

$$\tilde{L}_0(u) := 1 + \sum_{d=1}^{\infty} \frac{d}{5} w(\mathcal{O}_{h^3}\mathcal{O}_{h^{-1}})_{0,d} e^{du},$$
$$\tilde{L}_1(u) := 1 + \sum_{d=1}^{\infty} \frac{d}{5} w(\mathcal{O}_{h^2}\mathcal{O}_1)_{0,d} e^{du},$$
$$\tilde{L}_2(u) := 1 + \sum_{d=1}^{\infty} \frac{d}{5} w(\mathcal{O}_h\mathcal{O}_h)_{0,d} e^{du}. \tag{5.103}$$

Theorem 5.5.1

$$w(\mathcal{O}_{h^3}\mathcal{O}_{h^{-1}})_{0,d} = \frac{5}{d} \cdot \frac{(5d)!}{(d!)^5}. \tag{5.104}$$

Proof Note that the integral in the l.h.s. does not depend on order of integration if we take residues at $z_j = 0$, $\frac{z_{j-1}+z_{j+1}}{2}$ for $j = 1, 2, \ldots, d-1$ and at $z_j = 0$ for $j = 0, d$. Therefore, we integrate the l.h.s in ascending order of subscript j.

First, we integrate out the z_0 variable. By picking up the factors containing z_0, integration is done as follows:

$$\frac{1}{2\pi\sqrt{-1}} \oint_{C_{(0)}} \frac{dz_0}{z_0} 5 \prod_{i=0}^{4} (iz_0 + (5-i)z_1) \frac{1}{2z_1 - z_0 - z_2} = 5 \cdot 5! \frac{z_1^5}{2z_1 - z_2}. \tag{5.105}$$

Then we integrate the z_1 variable. Since the integrand is holomorphic at $z_1 = 0$, we only have to take the residue at $z_1 = \frac{z_2}{2}$:

$$5 \cdot 5! \frac{1}{2\pi\sqrt{-1}} \oint_{C_{(z_2 2)}} dz_1 5 z_2 \prod_{i=0}^{4} (iz_1 + (5-i)z_2) \frac{1}{2z_2 - z_1 - z_3} = 5 \cdot \frac{1}{2} \cdot \frac{10!}{(2!)^5} \frac{z_2^5}{\frac{3}{2}z_2 - z_3}. \tag{5.106}$$

Here, we used the identity:

$$\prod_{i=0}^{4} (i \cdot \frac{z_2}{2} + (5-i)z_2) = (z_2)^5 \prod_{i=0}^{4} \frac{10-i}{2} = (z_2)^5 \frac{10!}{5! \cdot 2^5}. \tag{5.107}$$

Integration of z_i ($i = 1, 2, \ldots, d-1$) goes in the same way. We only have to take the residue at $z_j = \frac{j}{j+1} z_{j+1}$. After finishing integration of z_{d-1}, what remains to do is the following integration.

$$5 \cdot \frac{1}{d} \cdot \frac{(5d)!}{(d!)^5} \frac{1}{2\pi\sqrt{-1}} \oint_{C_{(0)}} \frac{dz_d}{z_d}. \tag{5.108}$$

Hence we obtain the assertion of the theorem. □

This theorem leads us to the formula:

$$\tilde{w}_0(u) = \tilde{L}_0(u). \tag{5.109}$$

Hence we have reproduced the power series solution $\tilde{w}_0(u)$ of the differential equation of period integrals in Chap. 4 as the generating function of the intersection number $w(\mathcal{O}_{h^3}\mathcal{O}_{h^{-1}})_{0,d}$.

Theorem 5.5.2

$$\prod_{j=0}^{d}\left(\frac{1}{2\pi\sqrt{-1}}\oint_{C_j}\frac{dz_j}{(z_j)^5}\right)(z_0)^2 z_1\left(\prod_{j=1}^{d}e(z_{j-1},z_j)\right)\left(\prod_{j=1}^{d-1}\frac{1}{5z_j(2z_j-z_{j-1}-z_{j+1})}\right)\frac{1}{z_d}$$
$$=\frac{5}{d}\cdot\frac{(5d)!}{(d!)^5}\left(1-\frac{1}{d}+\sum_{j=1}^{5d}\frac{5}{j}-\sum_{j=1}^{d}\frac{5}{j}\right). \tag{5.110}$$

Proof In the same way as the proof of Theorem 5.5.1, we begin by integrating out the z_0 variable:

$$\frac{1}{2\pi\sqrt{-1}}\oint_{C_{(0)}}\frac{dz_0}{(z_0)^2}5\cdot z_1\prod_{i=0}^{4}(i\cdot z_0+(5-i)z_1)\frac{1}{2z_1-z_0-z_2}=5\cdot 5!\left((a_1)\frac{(z_1)^4}{2z_1-z_2}+\frac{(z_1)^5}{(2z_1-z_2)^2}\right), \tag{5.111}$$

where

$$a_1=\sum_{i=1}^{4}\frac{i}{5-i}. \tag{5.112}$$

In deriving (5.111), we used the following equality:

$$\frac{\partial}{\partial z_0}\left(\prod_{i=0}^{4}(i\cdot z_0+(5-i)z_1)\frac{1}{2z_1-z_0-z_2}\right)$$
$$=\left(\prod_{i=0}^{4}(i\cdot z_0+(5-i)z_1)\right)\frac{1}{2z_1-z_0-z_2}\left(\sum_{i=1}^{4}\frac{i}{i\cdot z_0+(5-i)z_1}+\frac{1}{2z_1-z_0-z_2}\right). \tag{5.113}$$

Since we have another z_1 factor in the integrand, it becomes holomorphic at $z_1=0$ after integration of z_0. Hence integration of the z_1 variable is done by taking the residue at $z_1=\frac{z_2}{2}$:

$$5\cdot 5!\frac{1}{2\pi\sqrt{-1}}\oint_{C_{(\frac{z_2}{2})}}dz_1\left((a_1)\frac{1}{2z_1-z_2}+\frac{z_1}{(2z_1-z_2)^2}\right)\times\prod_{i=0}^{4}(i\cdot z_1+(5-i)z_2)\frac{1}{2z_1-z_0-z_2}$$
$$=\frac{5}{2}\cdot\frac{10!}{(2!)^5}\left((a_2)\frac{(z_2)^5}{\frac{3}{2}z_2-z_3}+\frac{1}{4}\frac{(z_2)^6}{(\frac{3}{2}z_2-z_3)^2}\right). \tag{5.114}$$

Here, a_2 is given by

$$a_2 := a_1 + \frac{1}{2 \cdot 1} + \frac{1}{4}\sum_{i=1}^{4} \frac{2i}{10-i}. \tag{5.115}$$

In deriving (5.114), we used the following equality:

$$\frac{\partial}{\partial z_1}\left(z_1 (\prod_{i=0}^{4}(i \cdot z_1 + (5-i)z_2))\frac{1}{2z_2 - z_1 - z_3}\right)$$
$$= z_1\left(\prod_{i=0}^{4}(i \cdot z_1 + (5-i)z_2)\right)\frac{1}{2z_2 - z_1 - z_3} \times \left(\frac{1}{z_1} + \sum_{i=1}^{4}\frac{i}{(i \cdot z_1 + (5-i)z_2} + \frac{1}{2z_2 - z_1 - z_3}\right). \tag{5.116}$$

Integration of z_j $(j = 2, \ldots, d-2)$ goes in the same way. After finishing integration of z_{d-1}, the l.h.s. of (5.110) becomes

$$\frac{1}{d} \cdot 5 \cdot \frac{(5d)!}{(d!)^5}(a_d)\frac{1}{2\pi\sqrt{-1}}\oint_{C_{(0)}} \frac{dz_d}{z_d}, \tag{5.117}$$

where

$$a_d = \sum_{j=2}^{d}\frac{1}{j(j-1)} + \sum_{j=1}^{d}\frac{1}{j^2}\sum_{i=1}^{4}\frac{ji}{5j-i} = 1 - \frac{1}{d} + \sum_{j=1}^{5d}\frac{5}{j} - \sum_{j=1}^{d}\frac{5}{j}. \tag{5.118}$$

Integration of z_d immediately leads us to the assertion of the theorem. □

This theorem leads us to the formula:

$$\prod_{j=0}^{d}\left(\frac{1}{2\pi\sqrt{-1}}\oint_{C_j}\frac{dz_j}{(z_j)^5}\right)(z_0)^2(z_1 - z_0 + \frac{1}{d}z_0)\left(\prod_{j=1}^{d} e(z_{j-1}, z_j)\right) \times$$
$$\left(\prod_{j=1}^{d-1}\frac{1}{5z_j(2z_j - z_{j-1} - z_{j+1})}\right)\frac{1}{z_d} = \frac{5}{d}\frac{\partial}{\partial\varepsilon}\left(\frac{\prod_{j=1}^{5d}(j+5\varepsilon)}{\prod_{j=1}^{d}(j+\varepsilon)^5}\right)|_{\varepsilon=0}. \tag{5.119}$$

Proposition 5.5.1 *Let f be a positive integer that satisfies $1 \leq f \leq d-1$. Then the following equality holds:*

$$\prod_{j=0}^{d}\left(\frac{1}{2\pi\sqrt{-1}}\oint_{C_j}\frac{dz_j}{(z_j)^5}\right)(z_0)^2(2z_f - z_{f-1} - z_{f+1})\left(\prod_{j=1}^{d} e(z_{j-1}, z_j)\right)\times$$
$$\left(\prod_{j=1}^{d-1}\frac{1}{5z_j(2z_j - z_{j-1} - z_{j+1})}\right)\frac{1}{z_d}$$
$$= \frac{1}{5}w(\mathcal{O}_{h^2}\mathcal{O}_1)_{0,f}\cdot w(\mathcal{O}_{h^3}\mathcal{O}_{h^{-1}})_{0,d-f}. \tag{5.120}$$

Proof Since the integrand is holomorphic at $z_f = \frac{z_{f-1}+z_{f+1}}{2}$, the assertion follows from integration of z_i's in ascending order of the subscript i. □

By combining the above proposition with the following equality:

$$z_d + \sum_{f=1}^{d-1}(d-f)\cdot(2z_f - z_{f-1} - z_{f+1}) = d(z_1 - z_0) + z_0, \tag{5.121}$$

we can derive the following result from (5.119):

$$w(\mathcal{O}_{h^2}\mathcal{O}_1)_{0,d} + \sum_{f=1}^{d-1}(d-f)\frac{1}{5}w(\mathcal{O}_{h^2}\mathcal{O}_1)_{0,f}\cdot w(\mathcal{O}_{h^3}\mathcal{O}_{h^{-1}})_{0,d-f} = \frac{5}{d}\frac{\partial}{\partial\varepsilon}\left(\frac{\prod_{j=1}^{5d}(j+5\varepsilon)}{\prod_{j=1}^{d}(j+\varepsilon)^5}\right)\Big|_{\varepsilon=0}. \tag{5.122}$$

This is rewritten in terms of the generating function,

$$\begin{aligned}\left(\int^u dv\tilde{L}_1(v)\right)\tilde{L}_0(u) &:= \left(u + \sum_{d=1}^{\infty}\frac{w(\mathcal{O}_{h^2}\mathcal{O}_1)_{0,d}}{5}e^{du}\right)\left(1 + \sum_{d=1}^{\infty}\frac{dw(\mathcal{O}_{h^2}\mathcal{O}_1)_{0,d}}{5}e^{du}\right)\\ &= \left(u + \sum_{d=1}^{\infty}\frac{w(\mathcal{O}_{h^2}\mathcal{O}_1)_{0,d}}{5}e^{du}\right)\tilde{w}_0(u)\\ &= u\tilde{w}_0(u) + \tilde{w}_1(u).\end{aligned} \tag{5.123}$$

Here we used the formula (5.109). Therefore, we have reproduced the mirror map in Chap. 4 as the generating function of the intersection number $w(\mathcal{O}_{h^2}\mathcal{O}_1)_{0,d}$:

$$\begin{aligned}u + \sum_{d=1}^{\infty}\frac{w(\mathcal{O}_{h^2}\mathcal{O}_1)_{0,d}}{5}e^{du} &= u + \frac{\tilde{w}_1(u)}{\tilde{w}_0(u)} = t(u),\\ \tilde{L}_1(u) &= \frac{dt(u)}{du}.\end{aligned} \tag{5.124}$$

As for the intersection number $w(\mathcal{O}_h\mathcal{O}_h)_{0,d}$, we can prove the following theorem in the same way as the proof of Theorem 5.5.1 and Theorem 5.110.

Theorem 5.5.3

$$\prod_{j=0}^{d}\left(\frac{1}{2\pi\sqrt{-1}}\oint_{C_j}\frac{dz_j}{(z_j)^5}\right)(z_0)(z_1-z_0)(z_1-z_0+\frac{1}{d}z_0)\left(\prod_{j=1}^{d}e(z_{j-1},z_j)\right)\times$$
$$\left(\prod_{j=1}^{d-1}\frac{1}{5z_j(2z_j-z_{j-1}-z_{j+1})}\right)\frac{1}{z_d}=\frac{5}{2d}\frac{\partial^2}{\partial\varepsilon^2}\left(\frac{\prod_{j=1}^{5d}(j+5\varepsilon)}{\prod_{j=1}^{d}(j+\varepsilon)^5}\right)|_{\varepsilon=0}. \quad (5.125)$$

We leave the proof of this theorem to the reader as an exercise. By using some extensions of Proposition 5.5.1 and (5.121), we can derive the following equality:

$$w_2(u)=\tilde{L}_0(u)\int^{u}du_1\tilde{L}_1(u_1)\int^{u_1}du_2\tilde{L}_2(u_2),\Rightarrow\tilde{L}_2(u)=\frac{d}{du}\left(\frac{1}{\frac{dt(u)}{du}}\frac{d}{du}\left(\frac{w_2(u)}{\tilde{w}_0(u)}\right)\right). \quad (5.126)$$

Therefore, $\tilde{L}_2(u)$ is also written in terms of the period integral used in B-model computation. Though we omit detailed discussion for lack of space, we have the following theorem proved in [6].

Theorem 5.5.4

$$(\tilde{L}_0(u))^2(\tilde{L}_1(u))^2\tilde{L}_2(u)=\frac{1}{1-5^5e^u}. \quad (5.127)$$

Therefore, combining (5.109) and (5.124) with (5.127), we obtain

$$5\frac{\tilde{L}_2(u)}{\tilde{L}_1(u)}=\frac{1}{(\tilde{w}_0(u))^2}\frac{5}{1-5^5e^u}\left(\frac{1}{\frac{dt(u)}{du}}\right)^3. \quad (5.128)$$

If we invert the mirror map and substitute $u=u(t)$ into the above formula, we obtain,

$$5\frac{\tilde{L}_2(u(t))}{\tilde{L}_1(u(t))}=\frac{1}{(\tilde{w}_0(u(t)))^2}\frac{5}{1-5^5e^{u(t)}}\left(\frac{du(t)}{dt}\right)^3. \quad (5.129)$$

The r.h.s. is nothing but the formula obtained from the mirror symmetry hypothesis! If we write the r.h.s. as $5+\sum_{d=1}^{\infty}\langle\mathcal{O}_h(z_1)\mathcal{O}_h(z_2)\mathcal{O}_h(z_3)\rangle_d e^{dt}$, the above equation turns into

$$5\tilde{L}_2(u(t))\frac{du}{dt}=5+\sum_{d=1}^{\infty}\langle\mathcal{O}_h(z_1)\mathcal{O}_h(z_2)\mathcal{O}_h(z_3)\rangle_d e^{dt}. \quad (5.130)$$

By integrating out the both sides in t, we obtain

$$5\int^{u(t)} \tilde{L}_2(v)dv = 5u(t) + \sum_{d=1}^{\infty} w(\mathcal{O}_h\mathcal{O}_h)_{0,d}e^{du(t)}$$
$$= 5t + \sum_{d=1}^{\infty} \frac{\langle \mathcal{O}_h(z_1)\mathcal{O}_h(z_2)\mathcal{O}_h(z_3)\rangle_d}{d} e^{dt}, \qquad (5.131)$$

which confirms our numerical observation in (5.100).

Lastly, we explain the geometrical meaning of the coordinate transformation $u = u(t)$. For this purpose, we introduce P_d, which is a set of partitions of positive integer d:

$$P_d := \{\, \sigma_d = (d_1, d_2, \ldots, d_{l(\sigma_d)}) \mid \sum_{i=1}^{l(\sigma_d)} d_i = d,\ d_1 \geq d_2 \geq \cdots \geq d_{l(\sigma_d)} > 0 \,\}, \qquad (5.132)$$

where we denote a partition by σ_d. $l(\sigma_d)$ is the length of the partition σ_d. We also introduce the following notation:

$$\text{mul}(i, \sigma_d) = (\text{number of subscripts } j \text{that satisfy } d_j = i, \text{ where } \sigma_d = (d_1, \ldots, d_{l(\sigma_d)})). \quad (5.133)$$

With this set-up, we rewrite (5.131) into the form

$$5u + \sum_{d=1}^{\infty} w(\mathcal{O}_h\mathcal{O}_h)_{0,d}e^{du} = 5t(u) + \sum_{d=1}^{\infty} \frac{\langle \mathcal{O}_h(z_1)\mathcal{O}_h(z_2)\mathcal{O}_h(z_3)\rangle_d}{d} e^{dt(u)}. \quad (5.134)$$

By using (5.124), we obtain the following expansion:

$$e^{dt(u)} = \exp\left(du + d\sum_{f=1}^{\infty} \frac{w(\mathcal{O}_{h^2}\mathcal{O}_1)_{0,d}}{5} e^{fu} \right)$$
$$= \sum_{f=1}^{\infty} e^{(d+f)u} \sum_{\sigma_f \in P_f} d^{l(\sigma_f)} \prod_{i=1}^{l(\sigma_f)} \frac{w(\mathcal{O}_{h^2}\mathcal{O}_1)_{0,f_i}}{5} \prod_{j=1}^{f} \frac{1}{(\text{mul}(j, \sigma_f))!}. \quad (5.135)$$

The following equality is obtained from (5.134) and (5.135):

$$w(\mathcal{O}_h\mathcal{O}_h)_{0,d} = \frac{\langle \mathcal{O}_h(z_1)\mathcal{O}_h(z_2)\mathcal{O}_h(z_3)\rangle_d}{d} +$$
$$\sum_{f=1}^{d-1} \sum_{\sigma_f \in P_f} \frac{\langle \mathcal{O}_h(z_1)\mathcal{O}_h(z_2)\mathcal{O}_h(z_3)\rangle_{d-f}}{d-f} (d-f)^{l(\sigma_f)} \prod_{i=1}^{l(\sigma_f)} \frac{w(\mathcal{O}_{h^2}\mathcal{O}_1)_{0,f_i}}{5} \prod_{j=1}^{f} \frac{1}{(\text{mul}(j, \sigma_f))!} +$$
$$w(\mathcal{O}_{h^2}\mathcal{O}_1)_{0,d}. \qquad (5.136)$$

Then we introduce the following notation:

$$\langle \mathcal{O}_h \mathcal{O}_h \rangle_{0,d} := \frac{\langle \mathcal{O}_h(z_1)\mathcal{O}_h(z_2)\mathcal{O}_h(z_3)\rangle_d}{d},$$
$$\langle \mathcal{O}_h \mathcal{O}_h (\mathcal{O}_h)^{m+1} \rangle_{0,d} := \langle \mathcal{O}_h(z_1)\mathcal{O}_h(z_2)\mathcal{O}_h(z_3)\rangle_d d^m \ (d \geq 1,\ m \geq 0),$$
$$\langle \mathcal{O}_h \mathcal{O}_h \mathcal{O}_h \rangle_{0,0} := 5. \tag{5.137}$$

These are the so-called "genus 0 Gromov–Witten invariants" of a quintic hypersurface. With these notations, we rewrite (5.136) into the following form:

$$w(\mathcal{O}_h \mathcal{O}_h)_{0,d} = \langle \mathcal{O}_h \mathcal{O}_h \rangle_{0,d} +$$
$$\sum_{f=1}^{d-1} \sum_{\sigma_f \in P_f} \langle \mathcal{O}_h \mathcal{O}_h (\mathcal{O}_h)^{l(\sigma_f)} \rangle_{0,d-f} \prod_{i=1}^{l(\sigma_f)} \frac{w(\mathcal{O}_{h^2}\mathcal{O}_1)_{0,f_i}}{5} \prod_{j=1}^{f} \frac{1}{(\mathrm{mul}(j,\sigma_f))!} +$$
$$\langle \mathcal{O}_h \mathcal{O}_h \mathcal{O}_h \rangle_{0,0} \frac{w(\mathcal{O}_{h^2}\mathcal{O}_1)_{0,d}}{5}. \tag{5.138}$$

We assert that the above equation is the reflection of the decomposition of the moduli space,

$$CP^{5(d+1)-1} = \mathcal{M}_{CP^1}(CP^4, d) \sqcup \left(\bigsqcup_{f=1}^{d} \left(\mathcal{M}_{CP^1}(CP^4, d-f) \times CP^f \right) \right). \tag{5.139}$$

Of course, $w(\mathcal{O}_{h^a}\mathcal{O}_{h^b})_{0,d}$ is defined as an intersection number of $\overline{Mp}(CP^4, d)$. Though we have used another $\mathbf{C}^{\times}$ action in constructing $\overline{Mp}(CP^4, d)$, its spirit of compactification is the same as the one for $CP^{5(d+1)-1}$. On the other hand, Gromov–Witten invariants should be computed by using the uncompactified moduli space $\mathcal{M}_{CP^1}(CP^4, d)$. Therefore, we propose the following correspondence:

$$CP^{5(d+1)-1} \leftrightarrow w(\mathcal{O}_h \mathcal{O}_h)_{0,d},$$
$$\mathcal{M}_{CP^1}(CP^4, d) \leftrightarrow \langle \mathcal{O}_h \mathcal{O}_h \rangle_{0,d},$$
$$\mathcal{M}_{CP^1}(CP^4, d-f) \times CP^f \leftrightarrow \sum_{\sigma_f \in P_f} \langle \mathcal{O}_h \mathcal{O}_h (\mathcal{O}_h)^{l(\sigma_f)} \rangle_{0,d-f} \prod_{i=1}^{l(\sigma_f)} \frac{w(\mathcal{O}_{h^2}\mathcal{O}_1)_{0,f_i}}{5} \times$$
$$\prod_{j=1}^{f} \frac{1}{(\mathrm{mul}(j,\sigma_f))!} \quad (1 \leq f \leq d-1),$$
$$\mathcal{M}_{CP^1}(CP^4, 0) \times CP^d \leftrightarrow \langle \mathcal{O}_h \mathcal{O}_h \mathcal{O}_h \rangle_{0,0} \frac{w(\mathcal{O}_{h^2}\mathcal{O}_1)_{0,d}}{5}. \tag{5.140}$$

Then why does the partition of the positive integer f appear? We explain the reason in the following. $CP^f = \{(b_0 : b_1 : \cdots : b_f)\}$ was introduced to represent a degree d homogeneous polynomial in s and t:

$$b_0 t^f + b_1 t^{f-1} s + \cdots + b_f s^f . \tag{5.141}$$

But this is factorized uniquely into the following form up to multiplication of a non-zero constant:

$$\prod_{i=1}^{l(\sigma_f)} (\beta_i s - \alpha_i t)^{f_i}, \tag{5.142}$$

where $z_i := (\alpha_i : \beta_i)$'s represent mutually distinct points in CP^1. Note that z_i is associated with f_i in $\sigma_f = (f_1, \ldots, f_{l(\sigma_f)})$. Hence the partition σ_f appears. This fact leads us to the following decomposition of CP^f:

$$CP^f = \bigsqcup_{\sigma_f \in P_f} M_{0,l(\sigma_f)}(\sigma_f), \tag{5.143}$$

where

$$M_{0,l(\sigma_f)}(\sigma_f) := \{\ (z_1, z_2, \ldots, z_{l(\sigma_f)}) \mid z_i \in CP^1,\ z_i \neq z_j\ (i \neq j)\ \}/(\prod_{j=1}^{f} S_{\mathrm{mul}(j,\sigma_f)}). \tag{5.144}$$

In (5.144), S_d is the symmetric group of degree d ($S_0 = S_1 = \{1\}$) and $S_{\mathrm{mul}(j,\sigma_f)}$ acts as the permutation of z_i's whose corresponding f_i's equal j. This comes from the fact that the polynomial $\prod_{i=1}^{m}(\beta_i s - \alpha_i t)^j$ is invariant under permutation of $(\alpha_i : \beta_i)$'s. With this set-up, we can explain the geometrical meaning of the correspondence in (5.140) in the $1 \leq f \leq d-1$ case. Roughly speaking z_i in $M_{0,l(\sigma_f)}(\sigma_f)$ creates an additional insertion of $\mathcal{O}_h$ in the Gromov–Witten invariant $\langle \mathcal{O}_h \mathcal{O}_h \rangle_{0,d-f}$ and multiplication factor $\frac{w(\mathcal{O}_{h^2}\mathcal{O}_1)_{0,f_i}}{5}$. The factor $\prod_{j=1}^{f} \frac{1}{(\mathrm{mul}(j,\sigma_f))!}$ comes from division by $\prod_{j=1}^{f} S_{\mathrm{mul}(j,\sigma_f)}$ in the definition of $M_{0,l(\sigma_f)}(\sigma_f)$. Finally, we discuss the $f = d$ case. In this case, we have to consider the genus 0 and degree 0 Gromov–Witten invariant. But the general characteristics of Gromov–Witten invariants allow insertion of only three operators in this case. Hence in the decomposition of (5.143), only $M_{0,1}((d))$ contributes. This explains the correspondence in the $d = f$ case.

By rewriting (5.138) into the following form:

$$\langle \mathcal{O}_h \mathcal{O}_h \rangle_{0,d} = w(\mathcal{O}_h \mathcal{O}_h)_{0,d} - \sum_{f=1}^{d-1} \sum_{\sigma_f \in P_f} \langle \mathcal{O}_h \mathcal{O}_h (\mathcal{O}_h)^{l(\sigma_f)} \rangle_{0,d-f} \prod_{i=1}^{l(\sigma_f)} \frac{w(\mathcal{O}_{h^2} \mathcal{O}_1)_{0,f_i}}{5} \prod_{j=1}^{f} \frac{1}{(\mathrm{mul}(j, \sigma_f))!} - \langle \mathcal{O}_h \mathcal{O}_h \mathcal{O}_h \rangle_{0,0} \frac{w(\mathcal{O}_{h^2} \mathcal{O}_1)_{0,d}}{5}, \tag{5.145}$$

we can conclude that the coordinate change $u = u(t)$ is nothing but the process of subtracting unwanted contributions from boundary components added in compactification![2]

References

1. A.B. Givental. *Equivariant Gromov-Witten invariants.* Internat. Math. Res. Notices (1996), no. 13, 613–663
2. B.H. Lian, K. Liu, S.-T. Yau. *Mirror principle. I.* Asian J. Math. 1 (1997), no. 4, 729–763
3. I. Ciocan-Fontanine, B. Kim. *Wall-crossing in genus zero quasimap theory and mirror maps.* Algebr. Geom. 1 (2014), no. 4, 400–448
4. H. Saito. *Chow rings of* $\widetilde{Mp}_{0,2}(N,d)$ *and* $\overline{M}_{0,2}(\mathbf{P}^{N-1},d)$ *and Gromov–Witten invariants of projective hypersurfaces of degree* 1 *and* 2. Internat. J. Math. 28 (2017), no. 12, 1750090, 35 pp
5. R. Bott. *A residue formula for holomorphic vector-fields.* J. Differential Geometry 1 (1967) 311–330
6. M. Jinzenji. *Gauss–Manin system and the virtual structure constants.* Internat. J. Math. 13 (2002), no. 5, 445–477
7. M. Jinzenji. *Geometrical Proof of Generalized Mirror Transformation of Projective Hypersurfaces.* arXiv:1712.09819
8. M. Jinzenji. *Mirror map as generating function of intersection numbers: toric manifolds with two Kähler forms.* Comm. Math. Phys. 323 (2013), no. 2, 747–811

[2]Readers interested in the discussion in this part can refer to [7] for details.

Printed in the United States
By Bookmasters